CHEMICAL ENGINEERING METHODS AND TECHNOLOGY

HANDBOOK OF RESEARCH ON CHEMOINFORMATICS AND CHEMICAL ENGINEERING

CHEMICAL ENGINEERING METHODS AND TECHNOLOGY

Additional books in this series can be found on Nova's website under the Series tab.

Additional E-books in this series can be found on Nova's website under the E-book tab.

CHEMICAL ENGINEERING METHODS AND TECHNOLOGY

HANDBOOK OF RESEARCH ON CHEMOINFORMATICS AND CHEMICAL ENGINEERING

A. K. HAGHI
EDITOR

Nova Science Publishers, Inc.
New York

For permission to use material from this book please contact us:
Telephone 631-231-7269; Fax 631-231-8175
Web Site: http://www.novapublishers.com

LIBRARY OF CONGRESS CATALOGING-IN-PUBLICATION DATA

Handbook of research on chemoinformatics and chemical engineering / editor,
A.K. Haghi.
p. cm.
Includes index.
ISBN 978-1-62100-998-6 (softcover)
1. Chemical engineering. I. Haghi, A. K.
TP155.H276 2009
660--dc22

2010015608

Published by Nova Science Publishers, Inc. † New York

CONTENTS

PREFACE

The collection of topics in this book aims to reflect the diversity of recent advances in Chemistry and Chemoinformatics with a broad perspective which may be useful for scientists as well as for graduate students and engineers. This new book presents leading-edge research from around the world in this dynamic field.

Diverse topics published in this book are the original works of some of the brightest and well-known international scientists.

The book offers scope for academics, researchers, and engineering professionals to present their research and development works that have potential for applications in several disciplines of chemical engineering and science. Contributions ranged from new methods to novel applications of existing methods to gain understanding of the material and/or structural behavior of new and advanced systems.

Contributions are sought from many areas of science and chemical engineering in which advanced methods are used to formulate (model) and/or analyze the problem. In view of the different background of the expected audience, readers are requested to focus on the main ideas, and to highlight as much as possible the specific advantages that arise from applying modern ideas. A chapter may therefore be motivated by the specific problem, but just as well by the advanced method used which may be more generally applicable.

I would like to express my deep appreciation to all the authors for their outstanding contribution to this book and to express my sincere gratitude for their generosity. All the authors eagerly shared their experiences and expertise in this new book. Special thanks go to the referees for their valuable work.

In: Handbook of Research on Chemoinformatics ...
Editor: A.K. Haghi

ISBN: 978-1-62100-998-6

Chapter 1

ZrO2-TiN Composites

***Sedigheh Salehi*[1]*, Omer Van der Biest*[2] *and Jef Vleugels*[*3]**
Department of Metallurgy and Materials Engineering, Katholieke Universiteit Leuven, Kasteelpark Arenberg 44, B-3001 Heverlee, Belgium.

Abstract

1.75 mol % Y_2O_3-stabilized ZrO_2-based composites with 35-95 vol % TiN were fully densified by hot pressing for 1 hour at 1550°C under a load of 28 MPa. The TiN grain size was found to increase with increasing TiN content, resulting in a decreasing hardness and strength. The best mechanical properties, i.e., an indentation toughness of 5.9 $MPa.m^{1/2}$ in combination with a Vickers hardness of 14.7 GPa and an excellent bending strength of 1674 MPa were obtained for the composites with 40 vol % TiN. The active toughening mechanisms were identified and their contribution to the overall composite toughness are discussed. Transformation toughening was found to be the primary toughening mechanism in all investigated composites.

Keywords: Composites; Mechanical properties; Toughness; ZrO_2, TiN

1. Introduction

During the last decade, the applicability of zirconia to induce toughening by the stress-induced transformation of the tetragonal to monoclinic ZrO_2 phase in the stress field of propagating cracks, a phenomenon known as transformation toughening, [1,2] has been intensively investigated. Recent developments in zirconia composites are focused not only on the improvement of toughness, strength and hardness, but also on the possibility for mass production and manufacturing cost reduction. A successful approach is to incorporate

[1]Sedigheh.salehi@mtm.kuleuven.be
[2]Omer.VanderBiest@mtm.kuleuven.be
[*]Corresponding author. Jozef Vleugels, Phone: +32-16-321244; Fax: +32-16-321992
[3]Jozef.Vleugels@mtm.kuleuven.be

electrically conductive reinforcements such as TiB_2, [3] WC, [4] ZrB_2, [5] TiC, [6] TiCN [6] and TiN into the zirconia matrix. The incorporation of a certain content of these conductive reinforcements makes the composite electrically conductive enough to be machineable by electrical discharge machining. The goal of the present work is to investigate and evaluate the mechanical properties of ZrO_2-TiN composites with 35 up to 95 vol % TiN.

2. Experimental Procedure

Yttria-stabilised ZrO_2-TiN composites with 35-95 vol % TiN and 0.75 wt % Al_2O_3 were investigated. The commercial powders are yttria-free monoclinic ZrO_2 (Tosoh grade TZ-0, Tokyo, Japan; 27 nm grain size), 3 mol % yttria co-precipitated ZrO_2 (Tosoh grade TZ-3Y, Tokyo, Japan; 27 nm grain size) and jet-milled TiN (Kennametal, Victoria, USA; 1.03 μm grain size). The Y_2O_3-stabiliser content of the powder mixtures was adjusted by mixing the appropriate ratio of ZrO_2 starting powders. 0.75 wt % Al_2O_3 powder (Baikowski grade SM8, Annecy, France; 0.60 μm grain size) was added as a ZrO_2 grain growth inhibitor and sintering aid to all composite grades.

50 grams of powder was mixed on a multidirectional Turbula mixer in ethanol in a polyethylene container of 250 ml during 24 h at 60 rpm using 250 grams zirconia milling balls (ϕ = 3 mm). The dry powder mixture was inserted into a graphite die/punch set-up. After cold compression at 30 MPa, the samples were hot pressed in vacuum (~0.1 Pa) for 1 h under a mechanical load of 28 MPa at 1550°C with a heating rate of 50°C/min and a cooling rate of 10°C/min.

The density of the samples was measured in ethanol, according to the Archimedes method. The Vickers hardness (HV_{10}) was measured (Model FV-700, Future-Tech Corp., Tokyo, Japan) with an indentation load of 10 kg. The indentation toughness, K_{IC}, based on the crack length measurement of the radial crack pattern produced by Vickers HV_{10} indentations, was calculated according to the formula of Anstis et al.[7] The elastic modulus (E) was measured using the resonance frequency method.[8] The resonance frequency was measured by the impulse excitation technique (Model Grindo-Sonic, J. W. Lemmens N.V., Leuven, Belgium). The flexural strength at room temperature was measured in a 3-point bending test (Instron 4467, PA, USA) on rectangular samples ($45 \times 4.30 \times 1.44$ mm). All reported values are the mean and standard deviation of 5 measurements. Microstructural investigation was performed by scanning electron microscopy (SEM, XL-30FEG, FEI, Eindhoven, The Netherlands). X-ray diffraction (Seifert 3003 T/T, Ahrensburg, Germany) analysis was used to measure the relative monoclinic and tetragonal ZrO_2 phase content on polished and fractured composite surfaces.

3. Results and Discussion

3.1. Microstructures

The microstructural investigation of the samples revealed that all samples were fully densified since no pores were found on the cross-sections and the measured density of the composites is in the range of 99 to 100% of the theoretical density, calculated using a theoretical density for ZrO_2, TiN and Al_2O_3 of 6.05, 5.43 and 2.90 g/cm^3 respectively.

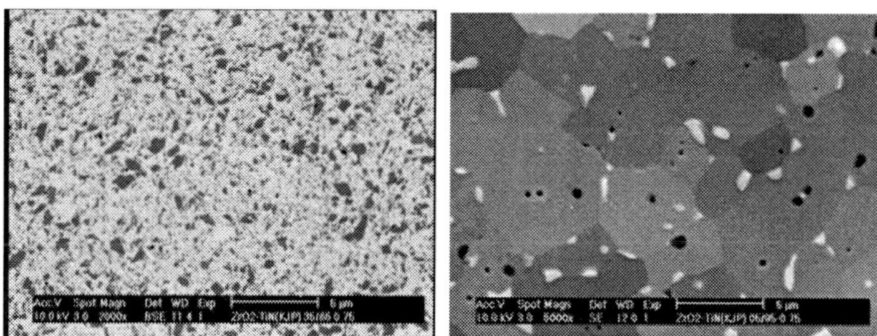

Figure 1.Backscattered SEM micrographs of polished 1.75 mol % yttria-stabilized ZrO_2-TiN-Al_2O_3 composites with 35 (left side) and 95 (right side) vol % TiN and 0.75 wt % Al_2O_3. Three phases can be differentiated; ZrO_2: bright, TiN: grey and Al_2O_3: black.

Moreover, homogeneous ZrO_2-TiN-Al_2O_3 microstructures were obtained, indicating that the powder mixing procedure was appropriate. The composite with 35 vol % TiN has the smallest TiN grain size, which is found to increase with increasing TiN content (Figure 1).

3.2. Mechanical Properties

In order to find the optimum yttria stabiliser content, the yttria content in ZrO_2-based composites with 40 vol % TiN and 0.75 wt % Al_2O_3 was varied from 2.5 down to 1.0 mol %. The resulting hardness and fracture toughness of the hot pressed composites is graphically presented in Fig. 2.

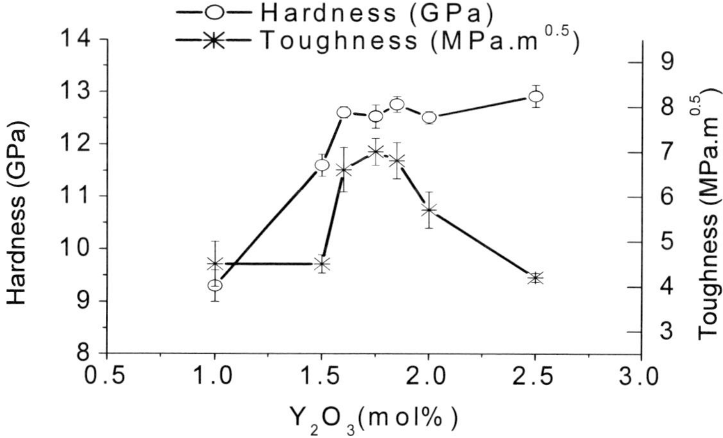

Figure 2. Hardness and toughness of hot pressed ZrO_2-TiN-Al_2O_3 (60/40/0.75) composites as function of the yttria stabiliser content. Indentation load: 5 kg)

The fracture toughness increases with increasing yttria content up to a maximum of 7 MPa m$^{1/2}$ at 1.75 mol % Y_2O_3 and decreases again upon addition of more stabiliser. This is completely in agreement with the toughness evolution in Y-TZP since the transformability and concomitant transformation toughness increases with decreasing yttria content, resulting in spontaneous transformation at too low stabiliser contents. The spontaneous transformation of the ZrO_2 phase is accompanied by microcracking, explaining the reduced hardness at yttria contents below 1.6 mol %. Because of the optimum toughness in combination with maximum hardness, an yttria content of 1.75 mol % was selected to investigate the influence of the TiN content.

The mechanical properties of the 1.75 mol % Y_2O_3-stabilised ZrO_2 with varying TiN content and 0.75 wt% Al_2O_3 are summarized in Table 1. The Vickers hardness and bending strength of the composites decreases with decreasing ZrO_2 content, what can be attributed to an increasing TiN grain size with increasing TiN content. The bending strength shows a maximum of 1674 MPa at 40 vol % TiN addition.

Table 1. Mechanical properties of 1.75 mol % Y_2O_3-stabilised ZrO_2-TiN composites with varying TiN content and 0.75 wt % Al_2O_3

TiN content [vol%]	ρ [g/cm^3]	E [GPa]	K_{IC} [MPa m$^{1/2}$]	HV_{10} [GPa]	3-point bending strength [MPa]
35	5.85	261 ±10	7.1 ± 0.2	13.75 ± 0.08	909 ± 140
40	5.81	290 ± 7	5.9 ± 0.1	13.65 ± 0.09	1674 ± 314
50	5.69	319 ± 5	6.3 ± 0.5	13.49 ± 0.26	1429 ± 61
60	5.60	344 ± 8	5.0 ± 0.3	13.55 ± 0.25	1464 ± 66
70	5.55	374 ± 6	4.7 ± 0.2	13.63 ± 0.16	1268 ± 87
80	5.49	402 ± 3	4.0 ± 0.5	13.36 ± 0.08	1307 ± 102
90	5.40	430 ± 6	4.8 ± 0.5	12.97 ± 0.18	916 ± 56
95	5.39	430 ± 15	3.3 ± 0.2	12.91 ± 0.29	678 ± 81

The contribution of transformation toughening to the overall fracture toughness of the composite can be calculated from the measured composite transformability, which is calculated from the measured ZrO_2 phase transformability multiplied by the ZrO_2 phase fraction in the composite [9]. The ZrO_2 phase transformability is defined as the difference in the percentage of monoclinic and tetragonal zirconia phase, obtained according to the formula of Toraya et al [9] from XRD patterns of smoothly polished and fractured surfaces. The m-ZrO_2 fraction measured on the fractured surface, the ZrO_2 phase transformability and the composite transformability are summarized in Table 2, allowing to calculate the actual transformation toughening contribution, assuming that a composite transformability of 63 % results in a toughness contribution of 6.6 MPa m$^{1/2}$.[4] Since no m-ZrO_2 was measured on the polished surfaces of all investigated composites, the ZrO_2 phase transformability is equal to the measured m-ZrO_2 content on the fractured surface.

The lower coefficient of thermal expansion of TiN ($\alpha_{0-1000°C}$ = 9.4×10^{-6}/°C) [10] compared to yttria-stabilized ZrO_2 ($\alpha_{0-1000°C}$ = 10 ×10^{-6}/°C) [11] will cause compressive residual stresses on the TiN phase. This residual compressive stress can provide an extra fracture toughening mechanism on the TiN phase that can be calculated as: [12]

$$\Delta K_{IC} = 2q\sqrt{\frac{2D}{\pi}}$$

Where K_{IC} is the critical stress intensity factor of the matrix, q is the local residual stress, and D is the length of the compressive stress zone, which in this case is the average particulate spacing of the zirconia particles. The local residual stress, q, can be calculated from the overall composite composition and the E modulus of the constituent phases according to the theoretical model proposed by Taya et al. [12] The presence of the Al_2O_3 phase was ignored for these calculations, whereas a Young's modulus of 471 and 200 GPa was used for TiN and ZrO_2 respectively. The ZrO_2 interparticle distance, D, was obtained according to the linear intercept method [13] using IMAGE-PRO software on SEM pictures. The calculated residual stress in the TiN phase, the measured interparticle distance and the concomitant toughness contribution due to the presence of the residual compressive stress in the TiN phase are summarized in Table 2. The tensile stresses generated in the ZrO_2 matrix are assumed to increase the transformability of the t-ZrO_2 phase and are incorporated in the experimentally measured composite transformability.

Table 2. Summary of the numerical values of the different toughening mechanisms for the yttria-stabilized ZrO_2-TiN-Al_2O_3 composites as a function of the TiN content, together with the values of the parameters needed for the calculations

TiN [vol%]	35	40	50	60	70	80	90	95
Zirconia phase transformability [%]	48	40	31	39	24	26	16	20
Composite transformability [%]	31	24	15	15	7	5	1	1
ΔK_{IC} transformation toughening [MPa m$^{1/2}$]	3.31	2.52	1.63	1.66	0.77	0.54	0.17	0.10
TiN Phase stress (q) [MPa]	214	194	157	122	89	57	28	13
ZrO_2 interparticle distance (D) [µm]	1.70	1.55	1.30	1.08	0.94	0.61	0.76	0.49
ΔK_{IC} thermal residual compressive stress on TiN [MPa m$^{1/2}$]	0.24	0.26	0.28	0.20	0.13	0.07	0.03	0.01
Secondary phase [vol %]	35	40	50	40	30	20	10	5
Matrix toughness [MPa m$^{1/2}$]	3.09	3.06	3.00	2.94	2.88	2.82	2.76	2.73
ΔK_{IC} crack deflection [MPa m$^{1/2}$]	1.12	1.15	1.03	0.97	0.91	0.83	0.95	0.61

Crack deflection was observed to be an active toughening mechanism in all composites, as illustrated in Fig. 3. The crack deflection model of Faber and Evans [14] predicts a toughness increase of 22 up to 37 % for 5 up to 50 % uniformly distributed spherical secondary phase particle addition. This model has been implemented to estimate the toughness contribution from crack deflection, as summarized in Table 2. A matrix toughness of 3.3 MPa.m$^{1/2}$ for non-transformable yttria-stabilized ZrO_2 [15] and 2.78 MPa.m$^{0.5}$ for TiN [16] was used.

The calculated toughening contribution from transformation toughening, compressive stresses and crack deflection (Table 2) are superimposed on the toughness of the non-transformable and stress free composite matrix in Fig. 4. The matrix toughness is calculated according to the rule of mixtures using a toughness of 3.3 MPa.m$^{1/2}$ for pure non-transformable yttria-stabilized ZrO_2 [15] and 2.78 MPa.m$^{0.5}$ for TiN[16]. The calculated total toughness is found to be in good agreement with the actually measured indentation toughness, revealing that it is possible to predict the fracture toughness of the composites with good accuracy. It must be mentioned that a possible interaction between the various toughening mechanisms has not been taken into account in the analysis. Both measured and calculated fracture toughness almost decreases linearly with the TiN content.

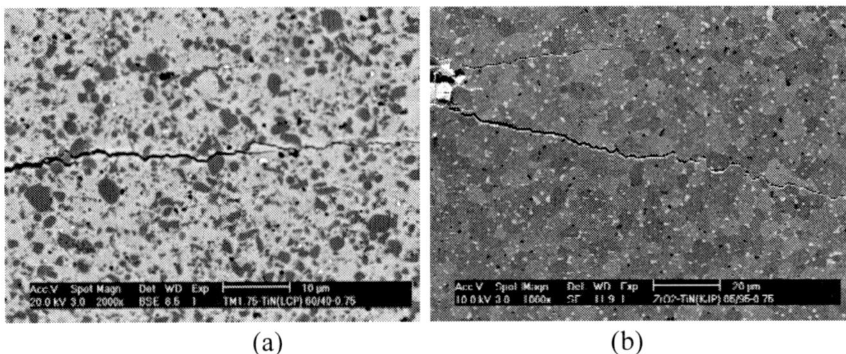

Figure 3. Crack propagation in yttria-stabilized ZrO2-TiN-Al2O3 composites with 40 (a) and 95 (b) vol % TiN. Three phases can be differentiated; ZrO2: bright, TiN: grey and Al2O3: black.

The major toughening mechanism is transformation toughening for the composites with up to 70 vol % TiN. Crack deflection by the secondary phase is an active toughening mechanism in all composites and becomes the primary toughening mechanism in the composites with 70 up to 95 vol % TiN. The toughening contribution from the compressive stress in the TiN phase is too low to have any significant effect.

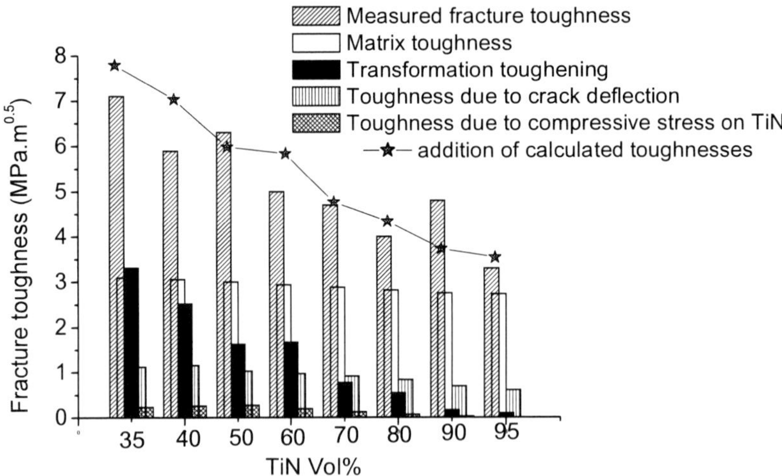

Figure 4. Measured and calculated toughness, together with the toughness contribution of the different toughening mechanisms.

4. Conclusions

Fully dense 1.75 mol % Y_2O_3 stabilised ZrO_2-TiN-Al_2O_3 composites with 0.75 wt % Al_2O_3 and a TiN content ranging from 35-95 vol % can be achieved by hot pressing at 1550°C for 1 hour. An optimum toughness was obtained with an yttria content of 1.75 mol % at a constant yttria content, the hardness, fracture toughness as well as flexural strength decrease almost linearly with TiN addition. The decreasing hardness and strength can be attributed to the larger TiN grain size with increasing TiN content, whereas the reduction in toughness is due to the decreasing contribution of transformation toughening. The primary toughening mechanism of the composites with up to 70 vol % TiN is transformation toughening, whereas crack deflection is the major toughening mechanism at higher TiN contents. The contribution of the residual compressive stress in the TiN phase is negligible. The best combination of hardness, toughness and strength is obtained for the ZrO_2-TiN composites with 40 vol % TiN addition.

ACKNOWLEDGMENTS

This work was performed in the framework of the MONCERAT project supported by the Commission of the European Communities within the FP6 Programme under project No. STRP 505541-1 and the Flemish Institute for the Promotion of Scientific and Technological Research in Industry (IWT) under contract number GBOU-IWT-010071.

REFERENCES

[1] Garvie, R. C., Hannink, R. H. J., Pascoe, R. T., *Nature.* 1975, 258, 703-704.
[2] Rühle, M. and Evans, A. G., Prog. *Mater. Sci.,* 1989, 33, 85-167.
[3] B.Basu, J. Vleugels, O. Van der Biest, Key Eng. *Mater.*, 2002, 206-213, 1177-1180.
[4] G. Anné, S. Put, K. Vanmeensel, D. Jiang, J. Vleugels, O. Van der Biest, *J. Europ. Ceram. Soc.,* 2005, 25, 55-63.
[5] B. Basu, J. Vleugels, O. Van der Biest, *J. Alloys and Comp.*, 2002, 334, 200-204.
[6] J. Vleugels, O. Van der Biest, *J. Amer. Ceram. Soc.*, 1999, 82, 2717 -2720.
[7] Anstis, G. R., Chantikul, P., Lawn, B. R. and Marshall, D. B., *J. Amer. Ceram. Soc,* 1981, 64, 533-538.
[8] ASTM Standard E 1876-99, *Test Method for Dynamic Young's Modulus, Shear Modulus, and Poisson's ratio for Advanced Ceramics by Impulse Excitation of Vibration,* ASTM Annual Book of Standards, PA, 1994.
[9] H. Toraya, M. Yoshimura and S. Somiya, *J. Amer. Ceram. Soc*, 1984, 67, 119-121.
[10] Peter J. *Hardro, selected material system and comparison of their predicted properties to H13 Tool steel*, Sep 1999, P5
[11] R. Morrell, *Handbook of properties of technical & engineering ceramics*, part1., 1985, 95-166.
[12] M. Taya, S. Hayashi, A. Kobayashi, H.S. Yoon, *J. Amer. Ceram. Soc,* 1990, 73, 1382-1391.
[13] M.I. Mendelson, *J. Amer. Ceram. Soc*, 1969, 52, 443-446.
[14] Faber, K. T. and Evans, A .G., *Acta Metall.,* 1983, 31, 565-576.
[15] Hannink, R. H. J., Kelly, P.M. and Muddle, B. C., *J. Amer. Ceram. Soc*, 2000, 83, 461-487.
[16] Toth LE., *Refractory materials, Transition metal carbides and nitrides*, Vol. 7, San Diego, Academic Press, 1971.

In: Handbook of Research on Chemoinformatics ... ISBN: 978-1-62100-998-6
Editor: A.K. Haghi

Chapter 2

PHOTOACTIVITY OF SUSPENDED AND IMMOBILIZED P-25 TiO_2 NANOPARTICLES ON EXAMPLE OF DIAZO CONGO RED DYE DEGRADATION

Lidija Ćurković* and Davor Ljubas
Faculty of Mechanical Engineering and Naval Architecture, University of Zagreb, Ivana Lučića 5, 10000 Zagreb, Croatia

ABSTRACT

This chapter of photocatalytic degradation of aqueous Congo Red dye (CR) solution has been investigated using two batch Pyrex glass photoreactor and artificial UV light (254 nm). For the first reactor A suspended TiO_2 (Degussa's P-25) nanoparticles were used, while for another reactor B TiO_2 (Degussa's P-25) nanoparticles were immobilized on inner surface of Pyrex glass. A spray coating technique was used for the immobilization of TiO_2 nanoparticles.

The experiments performed in both photoreactors indicate that the degradation of the Congo Red dye follows a pseudo-first order kinetics according to the Langmuir–Hinshelwood model. Photoactivity of immobilized TiO_2 layer for four times diluted CR solutions is almost equal to the photoactivity of the system with suspended photocatalyst. More than 80% of the initial concentration of CR is degraded in both reactors after 6 hours. But, looking at overall decrease of the CR concentration, in reactor A more CR molecules are degraded, since the initial concentration of the CR solution in reactor A is near 4 times higher.

Besides the photocatalytic degradation, adsorption process in reactor A on TiO_2 nanoparticles was also investigated and described.

Keywords: Photocatalytic degradation, UV radiation, TiO_2, Congo Red dye, Adsorption

*Corresponding author, e-mail: lidija.curkovic@fsb.hr, tel. +385-1-6168313

1. INTRODUCTION

Dyes that are in use today can be divided into several categories, based on their chemical nature: anionic or cationic and basic or reactive dyes. Azo dyes are the largest group of the synthetic colorants known and the most common group released in the environment. Wastewater containing dyes posses a serious hazard to living organisms and environment due to their potential high toxicity, carcinogenicity and mutagenicity effects (Brown and Devito, 1993; Alves de Lima et al., 2007). They are very difficult to be decolourised once released into the environment. Therefore, the removal of synthetic dyes with azo aromatic groups is a very important environmental task.

The development of new technologies for wastewater purification tends to reach complete destruction of the contaminants. Recently, advanced oxidation processes (AOPs) have been proposed as the alternative methods for water purification. Among AOPs, heterogeneous photocatalysis using TiO_2 as a photocatalyst is commonly used as destructive technology, starting with pioneering research of Fujishima and Honda (Fujishima and Honda, 1972). TiO_2 is a widely used photocatalyst due to its stability during UV light exposure, high chemical resistance, non-toxicity and low cost (Serpone and Pelizzeti, 1989). The photocatalytic efficiency of TiO_2 is dependent on its crystal structure, particle size and surface area. Titanium dioxide occurs in three crystalline polymorph forms: rutile (tetragonal), anatase (tetragonal) and brookite (orthorhombic) (Linsebigler et al.,1995). Rutile is known to be the most stable phase. Brookite and anatase are thermodynamically less stable than rutile. Generally, the TiO_2-based photocatalysts in anatase phase show usually the best photocatalytic efficiency. Degussa's P-25 TiO_2 product was identified in numerous of experiments as the best catalyst for heterogeneous photocatalytic oxidation of organic contaminants in water (Zumeta et al., 2003; Gumy et al., 2006; Kirchnerova et al., 2005; Zielinska et al., 2001) and it consists of two crystalline polymorph forms: anatase (ca. 75-80%) and rutile (ca. 20-25%).

Titanium dioxide can be used as a photocatalyst in various of applications, such as degradation of air pollutants (NO_x, aromates, chlorofluorocarbons) or aerosols (Hoffmann et al., 1995; Lackhoff and Niessner, 2002; Rajeshwar et al., 2008; Strini et al., 2005) and in the water purification processes, like degradation of various contaminants in waste water (Chen and Zhao, 1999; Franke and Franke, 1999; Fujishima et al, 2000; Gaya and Abdullah, 2008), removal of biofilms (Keskinen et al., 2006) or remediation of drinking and surface waters (Herrmann, 1999; Legrini et al., 1993; Ljubas, 2005). Additional benefit of using TiO_2 as the photocatalyst is a possibility for using the solar radiation for activation of the photodegradation process instead of using artificial UV sources (Kuo and Ho, 2006; Alfano et al., 2000; Melghit et al., 2009; Thu et al., 2005; Wu et al., 2007).

There are two possibilities of the practical use of TiO_2: as suspended nanoparticles in the liquid phase and as immobilized layer on the adequate substrate in a form of coating. The suspended TiO_2 nanoparticles are more effective due to the larger specific surface area of the catalyst available for reaction than TiO_2 coating. Two major problems are present when using suspended nanoparticles: the decrease of the light availability due to the scattering of UV light by nanoparticles and the separation of the photocatalyst nanoparticles after the treatment from the suspension. This could be solved by immobilizing TiO_2 nanoparticles as coatings on a solid substrate (e.g. reactor walls). There are many techniques for immobilization of TiO_2

catalysts onto the different types of substrates: dip coating, spin coating, spray coating, sol-gel, sputtering, electrophoretic deposition etc. (Byrne et el., 1998; Tennakone and Kottegoda, 1996 ; Hosseini et al, 2007).

Congo Red dye (CR), as a one of the most used azo dyes, especially to dye cotton, has been superseded by dyes more resistant to light and to washing. It is still used in histology to stain tissues for microscopic examination, and serves as an acid-base indicator, since it turns red in the presence of alkalis and blue when exposed to acids. Its chemical structure is the disodium salt of diphenyl diazo *bis*-1-naphthylaminesulphonic acid, is brownish red powder having absorbance maxima between 497.0 and 500.0 nm in aqueous medium. It is water soluble (reactive) secondary diazo dye and contains an azo (-N=N-) chromophore and an acidic auxochrome (sulfonate: $-SO_3H$) (which, respectively, gives and reinforces the coloration) associated with the benzene structure. It could also be used as a gamma-ray dosimeter since its coloration decays with the intensity of the irradiation (Parwate et al., 2007). The CR model solutions are interesting as the models of complex pollutants, too (Tapalad et al., 2008).

In this chapter, photocatalytic color removal of aqueous solution of CR dye has been described using commercial form of titanium dioxide from Degussa (P-25) in two forms: suspended in aqueous solution and immobilized on the inner surface of the reactor by spray methods.

2. Experimental Procedure

2.1. Materials and Apparatus

Titanium dioxide (TiO_2) catalyst from Degussa (P-25), with purity 99.9%, was used as received, without further modification. It is mostly in the anatase form (75-80% anatase and 20-25% rutile), nonporous, with a reactive surface (BET) area of 50 to 54 m^2 g^{-1}, corresponding to a mean particle size of around 30 nm (Alfano et al., 2000). Although the Degussa's TiO_2 photocatalyst from the series P-25 is well known and widely used, we have checked its properties using a transmission electron microscope EM 912 from ZEISS, its BET area using a surface area analyzer Autosorb from QUANTACHROME and X-ray diffraction with X-ray diffractometer D-5000 from SIEMENS.

It could be seen in the Figure 1 that the average diameter is mostly less than 100 nm, being in agreement with the reported average diameter of the P-25 TiO_2 particles around 30 nm.

The quantity of the sample of the P-25 TiO_2 for the specific area determination using BET isotherm model was 0.1167 g. Revealed result, using multipoint BET method and nitrogen as adsorbate, was 52,12 m^2/g, what is in a very good correlation with other research data (Zumeta et al., 2003).

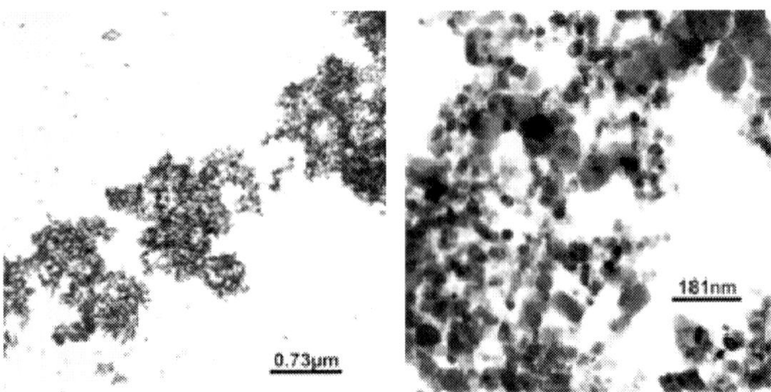

Figure 1. TEM visualisation of the Degussa's P-25 TiO_2.

Figure 2 showes the X-ray diffraction pattern of the Deggusa`s P-25 TiO_2. Results confirm that investigated TiO_2 nanoparticles consist of two crystalline polymorph forms: anatase and rutile.

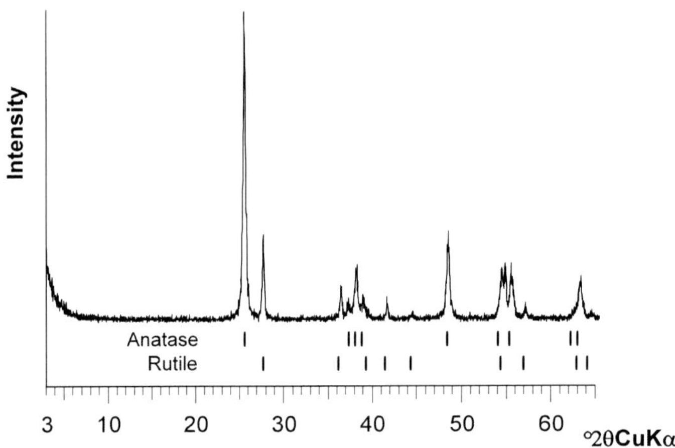

Figure 2. XRD diffractogram of the Degussa's P-25 TiO_2.

The structure of CR is illustrated in Fig. 3. CR was supplied by ACROS, as high purity biological stain, and used as a model compound without further purification.

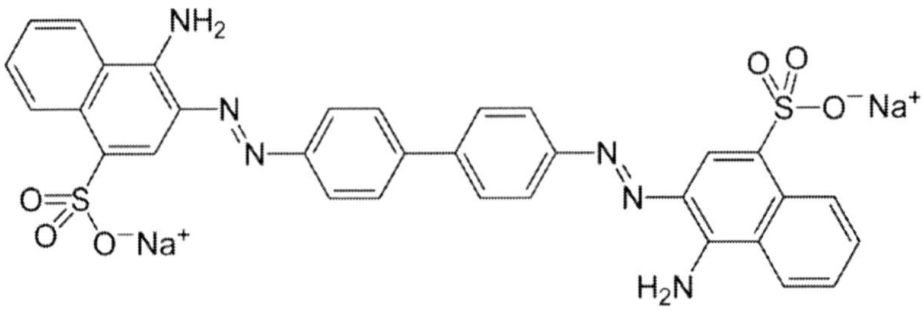

Figure 3. Molecular structure of Congo Red dye (formula: $C_{32}H_{22}N_6Na_2O_6S_2$; molecular weight: 696.66 g/mol).

All examining solutions were prepared using double distilled demineralized water.

2.2. Photocatalytic Decolorization

In order to investigate photocatalytic decolorization of diazo dye by TiO_2 (Degussa P25) nanoparticles, two Pyrex glass photoreactors were used: marked as photoreactor A and photoreactor B (Figure 4). In the photoreactor A TiO_2 nanoparticles were suspended in the aqueous solution of CR dye, while in the photoreactor B TiO_2 nanoparticles were immobilized on the inner surface of Pyrex glass. TiO_2 immobilization was done by a spray coating method. Both Pyrex glass photoreactor have volume of 0.2 L with 60 mm diameter. The radiation source were mercury UV-C lamps, λ_{max} = 254 nm, model Pen-Ray 90-0012-01, manufactured by UVP. The UV lamps were placed in the center of the reactor.

2.2.1. Photocatalytic Decolorization of Congo Red in the Photoreactor A

The removal of CR in photoreactors A and B (Figure 4) was investigated at the temperature (25±0.2) °C, with continuous purging with air (O_2). The reaction temperature was controlled by circulation of cooling water. The initial concentration of CR in the photoreactor A was 55 mg/L and concentrations of TiO_2 varied from 0.25 to 0.5 and 1.0 g/L.

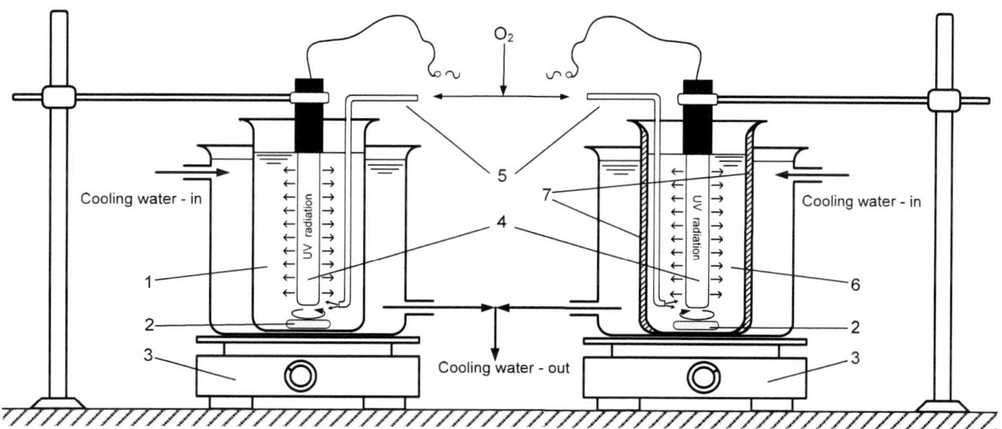

Figure 4. 1-The photoreactor A with suspended catalyst in aqueous solution of CR dye, 2-PTFE magnetic stir bar, 3-magnetic stirrer, 4-UV 254 nm mercury lamp, 5-glass tube for bubbling the solution with compressed air (O_2), 6-The photoreactor B, 7-spray TiO_2 coating on the inner surface of the reactor with examining solution of CR dye.

The mixtures of aqueous solution of CR and TiO_2 nanoparticles were stirred with the magnetic stirrers at a constant speed of 300 rpm. Before turning on the UV lamps, the solutions were placed in the dark, covered with aluminum foil and kept stirring for 15 min to reach the adsorption/desorption equilibrium of the suspension. The samples were taken out from the reactors for analysis at certain reaction intervals (60, 120, 240 and 360 min), centrifuged for 15 minutes for the samples from the Reactor A (SIGMA-LABORZENTRIFUGEN, Model 2-15) and remaining dye concentration was analyzed with UV-VIS spectrophotometer (HEWLETT PACKARD, Model HP 8430) at 498 nm using a 1 cm quartz cell. For each condition repetition tests were made to ensure reproducibility.

2.2.2. Photocatalytic Decolorization of Congo Red in Photoreactor B

The substrate (Pyrex glass) was cleaned by sonication in warm detergent and than rinsed in distilled water followed by sonication in distilled water for 15 min. Finally, the substrate was washed with the pure ethanol for 15 min. The inner surface of photoreactor B was coated by spraying a TiO_2 (Degussa's P-25)-methanol (1% w/v) suspension using a laboratory gun. The coating was dried using a blow drier and this procedure was repeated ten times. Then the coating was dried at 100°C for 4 h followed by calcination at 300 °C in a muffle furnace for 4 h.

The removal of CR in photoreactor B took place under the same conditions as those in the reactor A, except the initial concentration of CR. Due to high absorption of the CR solution on the wavelength of 254 nm, CR solution was diluted four times to ensure the irradiation of the reactor walls made of TiO_2.

3. Analysis of Decolorization of Diazo Congo Red Dye

3.1. Adsorption Isotherm Studies

The adsorption process of dissolved CR onto TiO_2 nanoparticles was investigated. Sorption experiments were performed in the above-described photoreactors. During the process of adsorption all the parameters were same as the parameters during the process of the photocatalytic degradation except that the UV lamp was switched off. The amount of TiO_2 in reactor A was constant (1 g TiO_2/L) and the concentrations of the CR varied from 2 to 55 mg/L. After 90 minutes of adsorption, the aqueous phase was separated from TiO_2 nanoparticles by centrifugation and concentration decrease of CR was determined by means of the absorbance measurement at 498 nm.

The amount of CR adsorbed on the TiO_2 nanoparticles was calculated from the following equation:

$$q_e = \frac{(c_0 - c_e) \cdot V}{m} \tag{1}$$

where q_e is the equilibrium CR concentration adsorbed on the TiO_2 nanoparticles (mg/g), V is the initial volume of CR solution used (L), m is the mass of TiO_2 used (g), c_0 is the initial concentration of CR in the solution (mg/L) and c_e is the equilibrium concentration of CR in the solution (mg/L).

The equilibrium distribution of CR between the sorbent (TiO_2 nanoparticles) and solution is important in determining the maximum sorption capacity. Several isotherm models are available to describe the equilibrium sorption distribution in which two models are used to fit experimental data: Langmuir and Freundlich models (Ćurković et al., 2001).

Results of adsorption research of CR dye onto TiO_2 nanoparticles are shown in Figure 5. These results are used for adsorption models evaluations according to Langmuir and Freundlich.

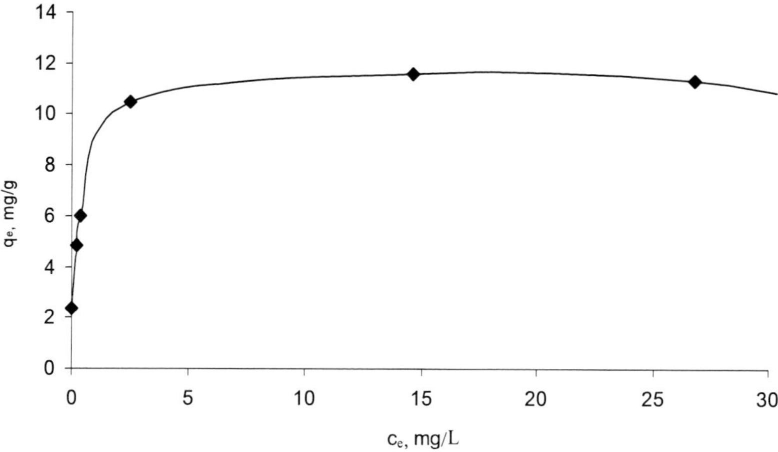

Figure 5. CR adsorption isotherm on the TiO_2 nanoparticles (1g TiO_2/L) at 298 K.

A basic assumption of the Langmuir theory is that the sorption takes place at specific homogeneous sites on the surface of the sorbent. It is then assumed that once a sorbate molecule occupies a site, no further sorption can take place at that site. The rate of sorption to the surface should be proportional to a driving force and area. The driving force is the concentration in the solution, and the area is the amount of bare surface. The Langmuir isotherm is expressed by:

$$q_e = \frac{q_m K_L c_e}{1 + K_L c_e} \tag{2}$$

The above equation can be rearranged to the following linear form:

$$\frac{1}{q_e} = \frac{1}{q_m} + \frac{1}{K_L q_m c_e} \tag{3}$$

where c_e is the equilibrium concentration (mg/L), q_e is the amount of CR sorbed (mg/g), q_m is the maximum amount of TiO_2 required to give a complete monolayer on the surface (mg/g), and K_L is an indication of the sorption capacity of the sorbent (L/mg). A plot of $1/q_e$ versus $1/c_e$ results in a straight line with a slope of ($1/K_L\ q_m$) and intercept of $1/q_m$ (not shown here). The Langmuir isotherm parameters were summarized in Table 1. The good agreement (R^2) with the experimental data suggests that the CR sorbed forms a monolayer coverage on the adsorbent surface of TiO_2 nanoparticles.

Freundlich's model presents an empirical sorption isotherm for non-ideal sorption on heterogeneous surfaces (McKay and Allen, 1980) as well as multilayer sorption and is expressed by the equation:

$$q_e = K_F \cdot c_e^{1/n} \tag{4}$$

where c_e is the equilibrium concentration of CR in the solution (mg/L), q_e is the equilibrium CR concentration adsorbed on the TiO_2 nanoparticles (mg/g), K_F (mg/g)/(mg/L)$^{1/n}$ and n are the Freundlich constants related to sorption capacity and sorption intensity, respectively.

The Langmuir isotherm data together with the Freundlich isotherm data for the investigated CR adsorption on TiO_2 nanoparticles are summarized in Table 1. Comparing the correlation coefficients listed in Table 1 we can draw the conclusion that the adsorption of CR on TiO_2 nanoparticles is better fitted with Langmuir isotherm equation under the concentration range studied. It indicates the monolayer coverage of TiO_2 nanoparticles by the CR dye.

Table 1. Langmuir and Freundlich isotherm parameters for the system CR- TiO_2 nanoparticles with TiO_2 load of 1 g/L

	Langmuir isotherm					Freundlich isotherm		
dye	q_m (mg/g)	K_L (L/mg)	R^2	R_L	γ_0 (mg/L)	n	K_F*	R^2
Congo Red	8.56	19.47	0.9151	0.02102	2.4	5.24	6.28	0.8567
				0.00963	5.3			
				0.00775	6.6			
				0.00387	13.2			
				0.00195	26.3			
				0.00128	40.0			
				0.00095	54.0			

*(mg/g)/(mg/L)$^{1/n}$

The essential characteristics of the Langmuir isotherm can be expressed in terms of a dimensionless constant separation factor or equilibrium parameter R_L (Eren and Acar, 2006):

$$R_L = 1/(1 + K_L\, c_o) \tag{6}$$

where R_L is dimensionless separation factor, c_0 is the initial concentration of CR (mg/L) and K_L is Langmuir constant, i.e. indication of the sorption capacity of the sorbent (L/mg). The R_L values (Eq. (6)) dictate favorable adsorption for $0 < R_L < 1$, as it is proposed. The parameter R_L indicates the shape of the isotherm accordingly Table 2. As shown in Table 1, R_L value decreases with the concentration and indicates favorable adsorption for CR dye on TiO_2 nanoparticles.

Table 2. The characteristics of separation factor R_L

R_L	Value type of isotherm
$R_L > 1$	Unfavorable
$R_L = 1$	Linear
$R_L = 0$	Irreversible
$0 < R_L < 1$	Favorable

3.2. Photocatalytic degradation

Figure 6A shows the time-dependent UV–VIS spectra of CR aqueous solution with concentration 55 mg/L as starting solution, during photocatalytic oxidation process. The spectrum of CR in the visible region has a maximum absorbance at 498 nm. The concentration of TiO_2 during the experiment was 1 g/L and the solution was irradiated with UV-radiation and continuously bubbled with air (O_2).

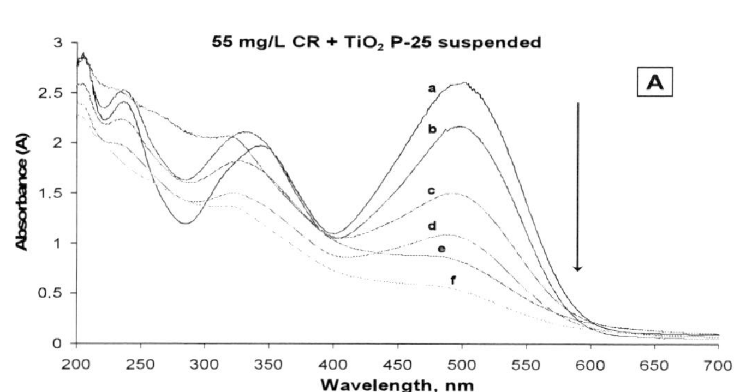

Figure 6. (A) Changes in the UV–VIS spectra of Congo red solution after the photocatalytic treatment in photoreactor A with UV-C radiation: (a) starting concentration of CR in aqueous solution 55 mg L^{-1}, (b) after 15 minutes of dark adsorption in the presence of 1 g L^{-1} TiO_2, (c) after 60 min of photocatalytic oxidation, (d) after 120 min of photocatalytic oxidation, (e) after 240 min of photocatalytic oxidation and (f) after 360 min of photocatalytic oxidation.

It is clear from figure 6 (A) and (B) that the intensity of the absorption peaks decreases with the time of UV-irradiation exposure, which implies that the CR is degraded. The initial pH value of the solution was 7.13 and after 360 min it decreased to the values between 5.1 and 5.95. Almost complete degradation of CR dye was observed after 360 min. Experiments performed "in the dark" confirmed that the effects of adsorption of CR on the surface of TiO_2 catalysts were negligible in the overall color removal in comparison to photocatalytic oxidation process (Figure 6A).

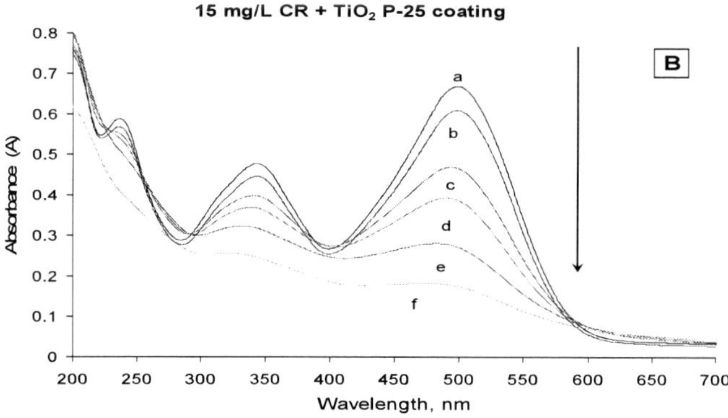

Figure 6. (B) Changes in the UV–VIS spectra of Congo red during photodegradation in photoreactor B: (a) 15 mg/L CR aqueous solution, (b) after 15 minutes of dark adsorption in the presence of spray TiO_2 coating, (c) after 60 min of photocatalytic oxidation, (d) after 120 min of photocatalytic oxidation, (e) after 240 min of photocatalytic oxidation and (f) after 360 min of photocatalytic oxidation and continuous purging with air (O_2).

3.3. Analysis of Experimental Conditions for Photoreactor A

Three set of experiments were performed in photoreactor A to compare CR degradation with and without TiO_2 catalysts (suspended nanoparticles). One set was performed with 55 mg CR/L exposed to different amount of TiO_2 nanoparticles but in absence of UV (0.25 g, 0.5 g and 1.0 g TiO_2 condition). The second set was performed by exposing 55 mg CR/L to UV-illumination in absence TiO_2 (the photolysis condition). The third set was performed by exposing 55 mg CR/L to different amount of TiO_2 nanoparticles (0.25 g, 0.5 g and 1.0 g TiO_2) in presence of UV-illumination (the photocatalysis condition). These results are presented in Figure 7A. The results with different amount of TiO_2 showed only that a small amount of CR dye (about 10% with 0.25 and 0.5 g TiO_2; about 22% with 1.0 g TiO_2) was adsorbed on the TiO_2 surface. The results of the photolysis and photocatalytic experiments showed that the photolysis reaction resulted in only 22% decrease in the CR concentration after 360 min while the CR was 82% degraded after 360 min in the case of the photocatalytic reaction. These results indicate that photocatalysis is more effective than direct photolysis for CR degradation.

Obtained results indicate that amount of TiO_2, as well as the duration of UV exposure, influences the degree of CR dye photodegradation. The photocatalytic efficiency for CR dye degradation increased with the increase of the amount of TiO_2. This observation can be explained by the photocatalyst optical properties as the main cause for the differences: the total active surface area increased with increasing catalyst dosage, but the overall photoactivity is limited by decreased availability of UV radiation due to shadowing and scattering of the UV light.

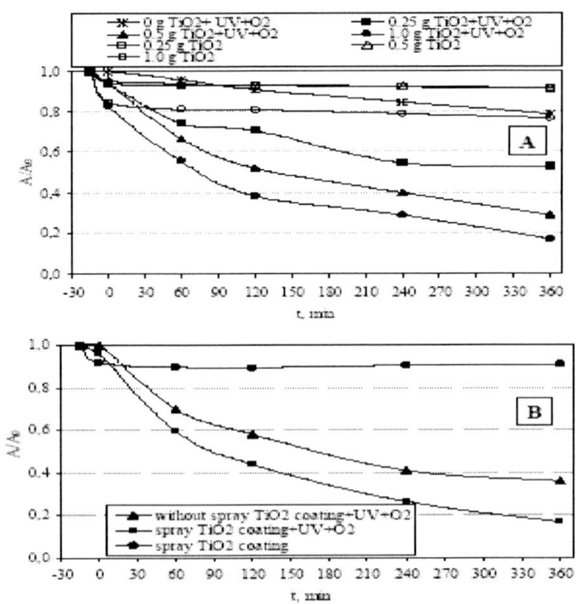

Figure 7. Time dependence of normalized absorbance for experiment performed in photoreactor A (A) and photoreactor B (B). Influence of the UV exposure time on the degree of CR dye photodegradation for different experimental conditions. (Time interval marked as "-15" to time "0" means 15 minutes of experiments "in the dark" (e.g. UV lamp was switched off), and the time interval from 0 to 360 minutes means that UV lamp was switched on).

The best result of photodegradation process with suspended TiO_2 is obtained using a TiO_2 concentration of 1.0 g/L + UV radiation + O_2 bubbling. Solely the UV-C radiation (predominantly by 254 nm) is capable for continuous degradation of CR. This degradation rate is lower (about 22% of the initial concentration of the CR is degraded after 360 min) than the degradation rate with presence of the photocatalyst, but it is still worth to consider, especially in the case of trying to design a real-scale reactor, because one of the major operational costs for photocatalyst removal (or photocatalyst loss) therewith could be avoided.

It is notable from the Figure 7 that decrease of normalized concentration of the CR solution is almost equal for both type of the reactor in the same time intervals. More than 80% of the initial concentration of CR is degraded in both reactors after 6 hours. But, looking at overall decrease of the CR, in reactor A more CR molecules are degraded, since the initial concentration of the CR solution in reactor A is near 4 times higher.

3.4. Analysis of Experimental Conditions for Photoreactor B

Three sets of experiments were performed in photoreactor B to compare CR degradation with and without spray TiO_2 coating as catalysts. The first set was performed with 15 mg CR/L exposed to spray TiO_2 coating but in absenceof UV. The second set was performed by exposing 15 mg CR/L to UV-illumination in the absence of spray TiO_2 coating (the photolysis condition). The third set was performed by exposing 15 mg CR/L to spray TiO_2 coating in presence of UV-illumination (the photocatalysis condition). These results are presented in Figure 7B. Results obtained with the first set of experiments showed that a small amount of CR (about 10%) was adsorbed on the TiO_2 surface. Results of the photolysis and photocatalytic experiments showed that the photolysis reaction resulted in about 65% decrease in the CR concentration after 360 min while the CR was about 85% removed after 360 min in the case of the photocatalytic reaction.

The initial concentration of CR in reactor B was near 4 times lower than in reactor A. The main reason for lower concentration in reactor B was the fact that in the area of the UV-C lamp radiation peak (254 nm) CR solution adsorbs the light strongly and therewith limits the UV light to reach the TiO_2 coating of the inner surface of the reactor.

The above results indicate that UV irradiation exposure of TiO_2 nanoparticles (in both case of suspended catalyst and immobilized catalyst) during 360 minutes continuously ensures the degradation of CR dye. It is in accordance with the theoretical background that explains energetics of electron processes in photocatalytic systems based on dispersed semiconductors (Hoffmann et al, 1995; Strini et al., 2005; Lackhoff and Niessner, 2002): when TiO_2 is continuously illuminated by light $\lambda < 400$ nm, electrons are continuously promoted from the valence band to the conduction band to give electron-hole pairs, as could be seen in Figure 8. The valence band potential is positive enough to generate hydroxyl radicals at the surface, and the conduction band potential is negative enough to reduce molecular oxygen. The hydroxyl radical is a powerful oxidizing agent and attacks organic pollutants that are present at or near the surface of TiO_2. It causes photooxidation process of the dye that might be explained partially according to the following reactions:

$$TiO_2 + h\nu \rightarrow TiO_2\left(e^-_{CB} + h^+_{VB}\right) \quad (7)$$

$$h^+_{VB} + H_2O_{ads} \rightarrow H^+ + {}^{\bullet}OH_{ads} \quad (8)$$

$$h^+_{VB} + OH^-_{ads} \rightarrow {}^{\bullet}OH_{ads} \quad (9)$$

$$e^-_{CB} + O_{2ads} \rightarrow {}^{\bullet}O^-_{ads} \quad (10)$$

$${}^{\bullet}OH_{ads} + \text{dye} \rightarrow \text{degradation of the dye} \quad (11)$$

$$h^+_{VB} + \text{dye} \rightarrow \text{dye}^{\bullet +} \rightarrow \text{degradation of the dye} \quad (12)$$

This mechanism is summarized in Fig. 8.

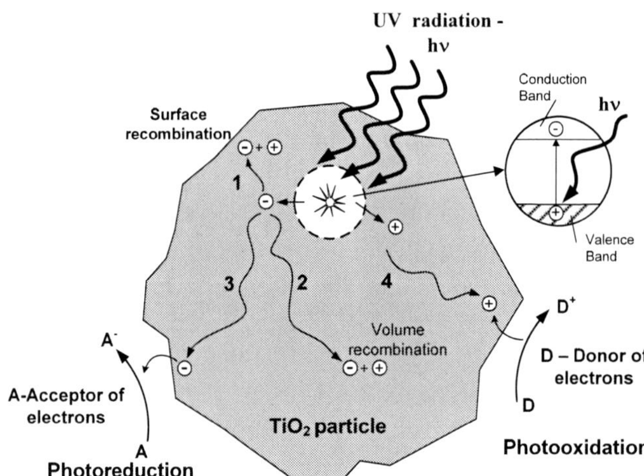

Figure 8. General mechanism of the photocatalytic process in TiO_2 water suspensions irradiated with UV light (Linsebigler et al., 1995).

3.5. Decolorization Kinetics

The photocatalytic decolorization of CR follows the Langmuir–Hinshelwood kinetics model given by the following equation (Al-Ekabi and Serpone, 1988):

$$-\frac{dc}{dt} = \frac{kKc}{1 + Kc} \quad (13)$$

where c is the CR concentration (mg/L) at time t (min), k the reaction rate constant (mg/L min), and K is the adsorption coefficient of CR (L/mg). After integration of Eq. (13) it is transformed to:

$$t = \frac{1}{Kk}\ln\left(\frac{c_0}{c}\right) + \frac{1}{k}\left(c_0 - c\right) \tag{14}$$

As the initial concentration (c) is is small, the second term on the right-hand side of Eq. (14) is relatively negligible (Konstantinou and Albanis, 2004). Therefore, Eq. (14) can be further simplified to give an apparent first-order equation:

$$-\ln\left(\frac{c_0}{c}\right) \cong kKt = k`t \tag{15}$$

where $k`$ is the pseudo-first order rate constant in min^{-1}.

The pseudo-first order rate constant $k`$ from Eq. (15) is evaluated through the linear regression of -ln ($c/c_0$) versus $t$ (Figure 9). The corresponding values of the pseudo-first order rate constant $k`$ as well as the correlation coefficient R^2 are given in Table 3.

The values of the pseudo-first order rate constants are higher for all amounts of TiO_2 in presence of UV irradiation comparing to the same amounts of TiO_2 without UV irradiation. Also, when introducing an increasing amount of TiO_2, an increase of the values of the pseudo-first order rate constants is observed (Table 3).

Obtained values of correlation coefficient R^2 (Table 3) indicate that the pseudo-first order rate is suitable for describing photodegradation process of CR photocatalytic degradation with TiO_2 as the photocatalyst.

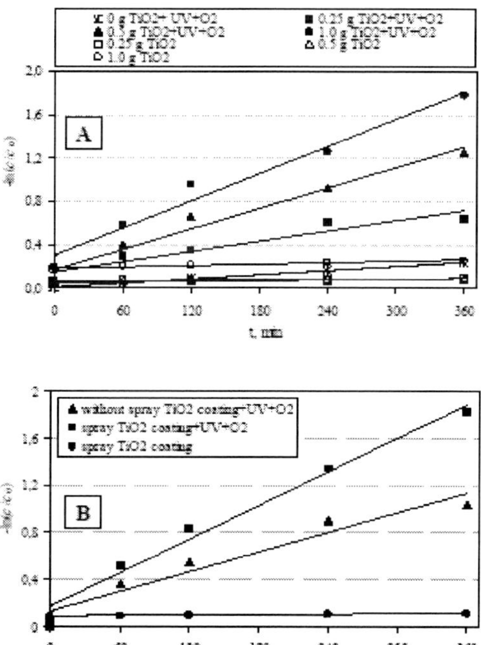

Figure 9. Plot of -ln (c/c_0) versus time of decolorization for different conditions in the batch photocatalytic reactor A (A) and reactor B (B).

Table 3. The pseudo-first order rate constant (k`) for decolorization process in different experimental conditions in the A (suspended TiO_2 in aqueous solution) and B (immobilized TiO_2) batch photocatalytic reactors

Reactor	Parameters in reactor	k`, min^{-1}	R^2
A	0 g TiO_2 + UV + O_2 + 55 mg CR/L	$6.6 \cdot 10^{-4}$	0.9961
	0.25 g TiO_2 + 55 mg CR/L	$0.8 \cdot 10^{-4}$	0.9688
	0.5 g TiO_2 + 55 mg CR/L	$1.1 \cdot 10^{-4}$	0.8837
	1.0 g TiO_2 + 55 mg CR/L	$2.4 \cdot 10^{-4}$	0.9097
	0.25 g TiO_2 + UV + O_2 + 55 mg CR/L	$15.8 \cdot 10^{-4}$	0.9015
	0.5 g TiO_2 + UV + O_2 + 55 mg CR/L	$31.3 \cdot 10^{-4}$	0.9668
	1.0 g TiO_2 + UV + O_2 + 55 mg CR/L	$41.9 \cdot 10^{-4}$	0.9725
B	spray TiO_2 coating + 15 mg CR/L	$0.7 \cdot 10^{-4}$	0.9208
	without TiO_2 coating+UV+O_2+ 15 mg CR/L	$27.8 \cdot 10^{-4}$	0.9292
	spray TiO_2 coating+UV+O_2+ 15 mg CR/L	$47.5 \cdot 10^{-4}$	0.9824

4. CONCLUSION

This chapter presents the results of photodegradation process of diazo Congo red dye (CR) aqueous solution with UV radiation and TiO_2 nanoparticles in two forms: suspended TiO_2 in aqueous solution and immobilized TiO_2 by spray method.

Based on the obtained results, the following conclusions can be drawn:

- For experiments performed with suspended TiO_2 nanoparticles (photoreactor A) it was observed that the degree of CR photodegradation increased with increasing the dose of the photocatalyst. The highest color removal rate was achieved using a TiO_2 dose of 1.0 g/L + UV radiation + O_2 bubbling.
- For experiments with immobilized TiO_2 (photoreactor B) it was observed that the degree of the CR photodegradation is higher for combination UV-C + TiO_2 coating than for only UV-C radiation.
- The decrease of normalized concentration of the CR solution is almost equal for both types of the reactor in the same time intervals. More than 80% of the initial concentration of CR, which was about four times lower in reactor B, is degraded in both reactors after 6 hours.
- Adsorption processes on the TiO_2 surface (photoreactor A) are better described using Langmuir's isotherm model than Freundlich's model.
- The values of R_L (dimensionless separation factor) for adsorption of CR on the surface of TiO_2 (photoreactor A) indicate that adsorption of CR is favorable on the adsorbent and the adsorption is more favorable at higher CR initial concentrations than at lower ones.
- The kinetics of photocatalytic degradation of CR in both photoreactors follows a pseudo-first order equation.
- The values of the pseudo-first order rate constants were found to increase with increase in photocatalyst concentration (photoreactor A).

ACKNOWLEDGMENTS

This chapter was supported by Ministry of Science, Education and Sports of Republic of Croatia within the framework of the Projects no.120-1253092-30 and 120-1201833-1789. Authors are grateful to the Research Centre Karlsruhe (Forschungszentrum Karlsruhe), Germany, for analytical assistance. D. Ljubas thanks to DAAD (German Academic Exchange Office) for the research fellowship.

REFERENCES

Alfano, O. M.; Bahnemann, D.; Cassano, A. E.; Dillert, R.; Goslich, R. *Photocatalysis in water environments using artificial and solar light*. Catal Today, 2000, 58, pp 199-230.

Alves de Lima, R. O.; Bazo, A. P.; Salvadori, D. M. F.; Rech, C. M. de P.; Oliveira, D. de A., Umbuzeiro, G. *Mutagenic and carcinogenic potential of a textile azo dye processing plant effluent that impacts a drinking water source*. Mutat Res - Gen Tox En, 2007, 626, pp 53-60.

Brown, M.A. and Devito S.C., *Predicting Azo-Dye Toxicity*. Crit Rev Env Sci Tec, 1993, 23(3), pp 249-324.

Byrne, J. A.; Eggins, B. R.; Brown, N. M. D.; Mckinney, B. and Rouse, M. *Immobilization of TiO2 powder for the treatment of polluted water*. Appl Catal B: Environ, 1998, 17, pp 25–36.

Chen, F. and J. Zhao, *Preparation and photocatalytic properties of a novel kind of loaded photocatalyst of TiO2/SiO2/γ - Fe2O3*. Catal Letters, 1999, 58, pp 246-247.

Ćurkovic, L., Cerjan-Stefanović S. and A. Rastovčan-Mioč, *Batch Pb2+ and Cu2+ removal by electric furnace slag*. Wat Res, 2001, 35: pp 3436-3440.

Eren, Z. and Acar, F.N., *Adsorption of Reactive Black 5 from an aqueous solution: equilibrium and kinetic studies*. Desalination, 2006, 194, pp 1-10.

Franke, R. and Franke C., *Model reactor for photocatalytic degradation of persistent chemicals in ponds and waste water*. Chemosphere, 1999, 39, pp 2651-2659.

Fujishima, A. and Honda K., *Electrochemical Photolysis of Water at a Semiconductor Electrode*. Nature, 1972, 238, pp 37-38.

Fujishima, A., Rao T.N., and Tryk D.A., *Titanium dioxide photocatalysis*. J Photochem Photobiol, A, 2000, 1, pp 1-21.

Gaya, U.I. and A.H. Abdullah, *Heterogeneous photocatalytic degradation of organic contaminants over titanium dioxide: A review of fundamentals, progress and problems*. Journal of Photochemistry and Photobiology C: Photochemistry Reviews, 2008. 9(1): p. 1-12.

Gumy, D.; Morais, C.; Bowen, P.; Pulgarin, C.; Giraldo, S.; Hajdu, R.; Kiwi, J. *Catalytic activity of commercial of TiO_2 powders for the abatement of the bacteria (E. coli) under solar simulated light: Influence of the isoelectric point*. Appl Catal B-Environ, 2006, 63, pp 76-84.

Al-Ekabi, H. and Serpone, N. *Kinetic studies in heterogeneous photocatalysis. 1. Photocatalytic degradation of chlorinated phenols in aerated aqueous solutions over TiO2 supported on a glass matrix*. J. Phys. Chem, 1988, 92, pp 5726-5731.

Herrmann, J.-M., *Heterogeneous photocatalysis: fundamentals and applications to the removal of various types of aqueous pollutants*. Catal Today, 1999, 53(1), pp 115-129.

Hoffmann, M. R.; Martin, S. T.;Choi, W. Y. and Bahnemann, D. W. *Environmental Applications of Semiconductor Photocatalysis*. Chem Rev, 1995, 95(1), pp 69-96.

Konstantinou I.K. *adn Albanis T.A.TiO2-assisted photocatalytic degradation of azo dyes in aqueous solution: kinetic and mechanistic investigations: A review* App Catal B-Environ, 2004,49, pp 1-14.

Tennakone K. and Kottegoda, I.R.M. *Photocatalytic mineralization of paraquat dissolved in water by TiO_2 supported on polythene and polypropylene films, J. Photochem. Photobiol.* A: Chem. 1996, 93 (1), pp 79-81.

Keskinen, H.; Mäkelä, J.; Aromaa, M.; Keskinen, J.; Areva, S.; Teixeira, C.; Rosenholm, J.; Pore, V.; Ritala, M.; Leskelä, M.; Raulio, M.; Salkinoja-Salonen, M.; Levänen, E.; Mäntylä, T. *Titania and titania-silver nanoparticle deposits made by Liquid Flame Spray and their functionality as photocatalyst for organic- and biofilm removal.* Catal Letters, 2006, 111, pp 127-132.

Kirchnerova, J.; Herrera Cohen, M. L.; Guy, C. and Klvana, D. *Photocatalytic oxidation of n-butanol under fluorescent visible light lamp over commercial TiO_2 (Hombicat UV100 and Degussa P25).* Appl Catal A-Gen, 2005, 282, pp 321-332.

Kuo, W.S. and P.H. Ho, *Solar photocatalytic decolorization of dyes in solution with TiO2 film.* Dyes Pigments, 2006, 71, pp 212-217.

Lackhoff, M. and Niessner R. *Photocatalytic atrazine degradation by synthetic minerals, atmospheric aerosols, and soil particles*. Environ Sci Technol, 2002, 36(24), pp 5342-5347.

Legrini, O., Oliveros, E. and Braun A.M., *Photochemical Processes for Water-Treatment.* Chem Rev, 1993, 93(2), pp 671-698.

Linsebigler, A.L., Lu, G.Q. and Yates, J.T., *Photocatalysis on Tio2 Surfaces - Principles, Mechanisms, and Selected Results*., Chem Rev, 1995, 95(3), pp 735-758.

Ljubas, D. *Solar photocatalysis - a possible step in drinking water treatment.* Energy, 2005, 30(10), pp 1699-1710.

McKay, G.; Allen S.J. *Surface mass transfer process using peat as an adsorbent for dyestuffs.* Can J Chem Eng, 1980, 58, pp 521-526.

Melghit, K., Al-Rubaei, M.S.; Al-Amri, I. *Photodegradation enhancement of Congo red aqueous solution using a mixture of $SnO_2{\cdot}xH_2O$ gel/ZnO powder.* J. Photochem. Photobiol. A: Chem., 2006, 181, pp 137-141.

Parwate, D.V.; Das Sarma, I.; Batra, R.J., *Preliminary feasibility study of congo red dye as a secondary dosimeter,* Radiation Meas, 2007, 42, pp 1527-1529.

Rajeshwar, K.; Osugi, M. E.; Chanmanee, W.; Chenthamarakshan, C. R.; Zanoni, M. V. B.; Kajitvichyanukul, P.; Krishnan-Ayer, R. *Heterogeneous photocatalytic treatment of organic dyes in air and aqueous media* J Photochem Photobiol, C, 2008, 9, pp 171-192.

Serpone, N.; Pelizzetti, E., Eds. *Photocatalysis : fundamentals and applications.* A Wiley-Interscience publication, John Wiley & Sons: New York ; Chichester; Brisbane; Toronto; Singapore, 1989.

Strini, A.; Cassese, S.; Schiavi, L. *Measurement of benzene, toluene, ethylbenzene and o-xylene gas phase photodegradation by titanium dioxide dispersed in cementitious materials using a mixed flow reactor.* App Catal B-Environ, 2005, 61, pp 90-97.

Tapalad, T.; Neramittagapong, A.; Neramittagapong, S.; Boonmee, M. *Degradation of Congo red dye by ozonation*, Chiang Mai J. Sci, 2008, 35(1), pp 63-68.

Thu, H.B.; Karkmaz, M.; Puzenat, E.; Guillard, Ch.; Herrmann, J.-M. *From the fundamentals of photocatalysis to its applications in environment protection and in solar purification of water in arid countries.* Res Chem Intermediat, 2005, 31, pp 449-461.

Wu, Y.; Liu, S.; Zuo, Y.; Li, J.; Wang, J. *Photodegradation of Some Dyes Over Ce/FSM-16 Catalyst Under Solar Light.* Catal Letters, 2007, 119, pp 245-251.

Zielinska, B.; Grzechulska, J.; Grzmil, B.; Morawski, A. W. *Photocatalytic degradation of Reactive Black 5: A comparison between TiO_2-Tytanpol A11 and TiO_2-Degussa P25 photocatalysts.* Appl Catal B-Environ, 2001, 35, pp L1-L7.

Zumeta, I.; Espinosa, R.; Ayllón, J. A.; Domčnech, X.; Rodríguez-Clemente, R.; Vigil, E. *Comparative study of nanocrystalline TiO_2 photoelectrodes based on characteristics of nanopowder used.* Sol Energ Mat Sol C, 2003, 76, pp 15-24.

In: Handbook of Research on Chemoinformatics ...
Editor: A.K. Haghi
ISBN: 978-1-62100-998-6

Chapter 3

CHARACTERIZATION AND MONITORING OF DISSOLUTION OF METAL STRINGS

Lidija Ćurković*[1] and Iva Rezić[2]

[1]Department of Materials, Faculty of Mechanical Engineering and Naval Architecture, University of Zagreb, Croatia

[2]Department of Applied Chemistry, Faculty of Textile Technology, University of Zagreb, Croatia

ABSTRACT

Metal strings which are coming into direct and prolonged contact with human skin are very important part of many musical instruments. Since metal ions may provoke skin allergic reactions, chemical composition and microstructure of metal strings directly influences human health.

In this chapter, monitoring of dissolution of electric guitar strings (E6, A5, D4, g3, h2 and e1) in artificial sweat solution were investigated.

Prior to performing these experiments it was very important to carry out the detailed characterization of investigated strings. In order to characterize the electric guitar strings different analytical techniques was used: inductively coupled plasma optical emission spectroscopy (ICP-OES), atomic absorption spectrophotometry (AAS), optical microscopy (OM) and a scanning electron microscopy with an energy dispersive X-Ray spectrometer (SEM-EDS). Obtained results indicate that electric guitar strings (E6, A5 and D4) consist of two separate parts: Sn-plated steel core wire which is hexagonal in cross section and Ni-plated steel wrap which is round in cross section. Furthermore, the electric guitar strings (g3, h2 and e1) consist of only one wire: Sn-plated steel core wire which is round in cross section.

Dissolution of electric guitar strings (E6, A5, D4, g3, h2 and e1) in artificial sweat solution was studied as a function of immersion time at the temperature of 37 °C in compliance to the EN 1811:1999 standard test procedure. From the amounts of Ni^{2+}, Mn^{2+}, Si^{4+}, Sn^{2+} and Fe^{3+} ions eluted in artificial sweat solution, a degree of dissolution χ_i for each component was calculated according to the equation $\chi_i = A/B$, where A and B are respectively the amounts of elements eluted in artificial sweat solution and the amounts of elements in the

* Corresponding author, e-mail: lidija.curkovic@fsb.hr, tel. +385-1-6168313

untreated material. The monitoring of the amount of metal ions eluted in artificial sweat solution was carried out by means of inductively coupled plasma optical emission spectroscopy (ICP-OES).

As testing samples, strings for electric guitar were chosen due to a content of nickel, the mostly known skin allergen. The results have proven that the samples investigated should not present a thread to human health, since the threshold of 0.5 μg Ni^{2+}/ cm^2 per week has not been over passed.

Keywords: metal strings, dissolution, artificial sweat, ICP-OES, AAS, OM, SEM-EDS.

1. Introduction

One of the earliest applications of metal fibers is their combination with natural animal fibers for manufacturing of metal strings for musical instruments. In recent time, animal fibers in the core of strings are replaced with polymer materials like nylon, Kevlar and carbon fibers, or with other metal wires (steel, phosphorus bronze, nickel, copper, gold, silver, tungsten and other metals). Characterization of such materials is extremely important because metal ions present in strings may have a negative impact on human health. One of the most common problems is allergic dermatitis to nickel and other metals.

According to a rough estimation, guitar is today the most popular instrument in the world [1-7]. The sound is produced by transversal vibration of the string, and transported by longitudinal vibration of sound particles. The quality of the sound strongly depends on the material used for the string manufacture, as well as on the resonant body of the instrument. Strings for instruments can be composed of one or more metal/alloy wires. Each wire usually consists of a core which is made of different natural or metal/alloy fibers, and the outer wire-wrap which is sometimes made of gold (violin strings) or silver (guitar strings) for prevention of string corrosion and allergic reactions on musician's skin. Metal strings are very sustainable to corrosion and oxidation due to salts and moisture in which they come into the contact. In order to prevent such undesirable processes, many manufacturers produce strings which are additional covered by polymer material.

String instruments are the most popular instruments in the world, whose sound quality is strongly influenced by their chemical and phase composition. Therefore the string's corrosion process provoked by human sweat has big influence on the quality of the performance. Secondly, corrosion process may cause skin allergic reactions [8]. Dermal hypersensitivity to metal is common in western world. It affects about 20% of female and 6 % of male population. Contact dermatitis is not caused by nickel itself but by nickel salts which are formed under the effect of perspiration in contact with an object, e.g. a guitar string. This phenomenon is always accompanied by corrosion of the object.

It is estimated that 86.5% of musicians playing string instruments have different skin problems. The occurrence of the problems depends on the time of the status of the musician [9], stress [10] and the time of the exposure [11]. The chronicle skin damages may lead to infections like *herpes labialis* [12]. Therefore the usage of strings with protective polymer covering is often recommended [13-15]. According to the British standard, the correlation between skin reaction and the release of nickel from alloys in the contact with artificial sweat is confirmed: the estimated threshold of nickel release, above which the allergy is triggered, is

0.5 μg Ni^{2+} cm^{-2} $week^{-1}$ [16]. Nevertheless some toxicologists recommend the threshold values of 0.2 μg cm^{-2} $week^{-1}$. Recent investigations on nickel release showed that the release of nickel from current 1- and 2 euro coins exceeds the acceptable value for prolonged contact with human skin by a factor of 240-320 [16,17]. Similar results were obtained for the euro coins' investigations of Fournier et al. [18, 19] and Nestle et al. [20]. Secondly, Gou et al. [21] reported nickel enrichment on both segments of 1 euro coins.

Nickel is considered to be the most important allergen today [22]. It is known that people exposed to nickel containing coins can be sensitized to nickel during their work [23]. Some working groups like bingo-hall workers and bank cashiers are considered to be at higher risks [24]. Therefore the nickel release investigations are reaching bigger proportions. I. Milošev et al. [25] monitored release of copper, nickel and zinc ions from their alloys. They have proved that the concentration of dissolved nickel exceeds the threshold concentration of nickel above which the allergy is triggered.

Since contact dermatitis can be provoked by nickel, several investigations have focused on the release of nickel from various stainless steel and Nickel alloys in the artificial sweat [26-32]. These studies showed that the dissolved amount of nickel from stainless steel was less than 0.5 $\mu g cm^{-2}$ $week^{-1}$. Chiba et al. [30] investigated dissolution of nickel, chromium and iron from stainless steels in artificial sweat. They have found out that the ratio of the three dissolved metal ions was not consistent with the composition of the investigated samples. Milošev and Kosec [29] reported amount of released copper, nickel and zinc from nickel-silver alloy after immersion in artificial sweat solution for 30 days. The authors reported that the mechanism of nickel release is the function of the composition of the artificial sweat solution. Amount of dissolved nickel was prevailed over dissolved of copper and zinc. Colin et al. [31] found out that nickel release did not necessary increase with the nickel content of the alloy. Our previous findings have proved that nickel present in everyday items like coins may induce the serious nickel allergy [32]. Therefore it should be monitored in different metal/alloys samples.

Nickel plated steel is one of the most popular electric guitar string due to its balance tone. However, there are neither data nor evidence of nickel and other metal ions harmfulness, concerning the release from guitar strings. The objective of the study presented in this chapter is, therefore, to evaluate the metal ions release from electric guitar strings after exposure to artificial sweat solution. The study of elution of nickel ion from electric guitar strings in artificial sweat allows prediction of reaction with the skin and possible allergenic reactions. Secondly, corrosion influences the sound quality significantly and may destroy the string after prolonged exposure to the oxidizing agents in the corrosive environment.

In this chapter, simultaneous determination of elution of nickel as well as other metal ions from electric guitar strings was performed by inductively coupled plasma-optical emission spectroscopy (ICP-OES). The ICP-OES is an excellent choice for simultaneous multi-elemental determination of elements in different samples due to low detection limits, large analytical working range and high precision [33]. In many cases the axially viewed system is favorized for its sensitivity [34], lower limits of detection [35] and precision, dynamic range and matrix effects on line signals [36]. This method is optimal for determination of metals present in everyday items [37].

The characterization and monitoring of dissolution of electric guitar strings marked as E6, A5, D4, g3, h2 and e1 are presented in this chapter. The special focus was to monitor nickel, which is especially harmful allergenic element. These investigations were performed due to their importance for protection of human health.

2. EXPERIMENTAL PROCEDURE

2.1. Characterization of Electric Guitar Strings

Investigated samples were electric guitar strings marked as E6, A5, D4, g3, h2 and e1. Strings are produced by D'Addario Company, USA and are used as the E6, A5, D4, g3, h2 and e1 strings for the electrical guitar.

Vickers hardness measurements HV0.1 was performed using indentation load of 0.9807 N for 15 s. Indentation tests were carried out under laboratory conditions on an Instron, Wilson-Wolpert Tukon 2100B, a micro Vickers tester. Before performing the hardness measurements, all strings were prepared by the standard metallographic technique. The samples were cut, embedded in resin DuroFix-2 Kit, ground and then polished up to 1 μm with a diamond paste until a mirror-like surface was achieved. Electric guitar strings were characterized by optical microscopy (OM) and scanning electron microscopy (SEM).

After chemically etching by immersion in Nital (2 wt % HNO_3 in ethanol) microstructure of strings was analyzed by optical microscope Olympus GX 51.

Chemical compositions of coating on electric guitar strings were determined by means of SEM (Tescan Vega TS5136LS) equipped with a EDX detector.

Chemical composition of core and wrap were determined by means of inductively coupled plasma optical emission spectroscopy (ICP-OES), an axially viewed Thermo Elemental "IRIS Intrepid II" XSP, Duo, (Thermo Electron Corporation). The instrument was equipped with a standard one piece extended torch with a quartz injector tube, a cyclone spray chamber and a concentric nebulizer.

Major and minor components from different sample parts were checked by flame atomic absorption spectrophotometer (AAS), AA-6800, Shimadzu.

2.2. Monitoring of Metal Ions Dissolution

Monitoring of metal ions dissolution was performed in artificial sweat solution prepared according to the EN 1811:1999. All the reagents used for this research work were of p.a. grade, supplied by Merck, Darmstadt, Germany. Ultra pure nitric acid supplied by Merck, Darmstadt, Germany and ultra pure water produced by Gen Pure TKA-LAB-HP, Nirosta, was applied for the ICP-OES analysis.

The strings E6, A5, D4, g3, h2 and e1 were simultaneously put into a sealed polypropylene (PP) tube with 10 ml of artificial sweat. The static corrosion tests were carried out at the temperature of 37 °C from 7 days (168 h) to 28 days (672 h). In parallel with the corrosion testing, a blind test was performed. The measurements were conducted after 7 days (168 h), 14 days (336 h), 21 days (504 h) and 28 days (672 h) of immersion. After the planned exposure time, the specimens were removed from the tubes, rinsed with distilled water, dried in a warm oven heated at 100 °C. The determination of the amounts of Ni^{2+}, Sn^{2+},

Si^{4+}, Mn^{2+} and Fe^{3+} ions released in the corrosive solutions was carried out by means of inductively coupled plasma optical emission spectroscopy (ICP-OES).

From the amounts of Ni^{2+}, Sn^{2+}, Si^{4+}, Mn^{2+} and Fe^{3+} ions eluted in artificial sweat solution, a degree of dissolution χ_i for each metal ion was calculated according to equation $\chi_i = A/B$, where A and B are respectively the amounts of elements eluted in the artificial sweat solution and the amounts of elements in the untreated material.

3. Results and Discussion

3.1 Results of Characterization of Electric Guitar Strings

Optical micrographs of polished cross-sections (Figure 1) show that strings marked as E6, A5 and D4 consist of two wires: a core and wrap. Both wires (core and wrap) have coating (Fig. 2B1, 2B2, 2C1 and 2C2 and Fig. 3). Core is hexagonal in cross section while wrap is round in it's cross section (Fig. 1). Shape of cross section of E6, A5 and D4 strings is the same and essential difference is only in diameter of the strings. Diameter of E6, A5 and D4 strings is 1.07 mm, 0.81 mm, and 0.61 mm, respectively (Fig. 1).

Strings marked as g3, h2 and e1 consist of only one wire: a core. Their cores are round in cross section and have coatings (Fig. 1, 2B1 and 2B2). Shapes of cross sections of g3, h2 and e1 string are round and essential difference is only in the diameter of the strings. Diameter of g3, h2 and e1 strings is 0.41 mm, 0.28 mm and 0.23 mm, respectively (Fig. 1).

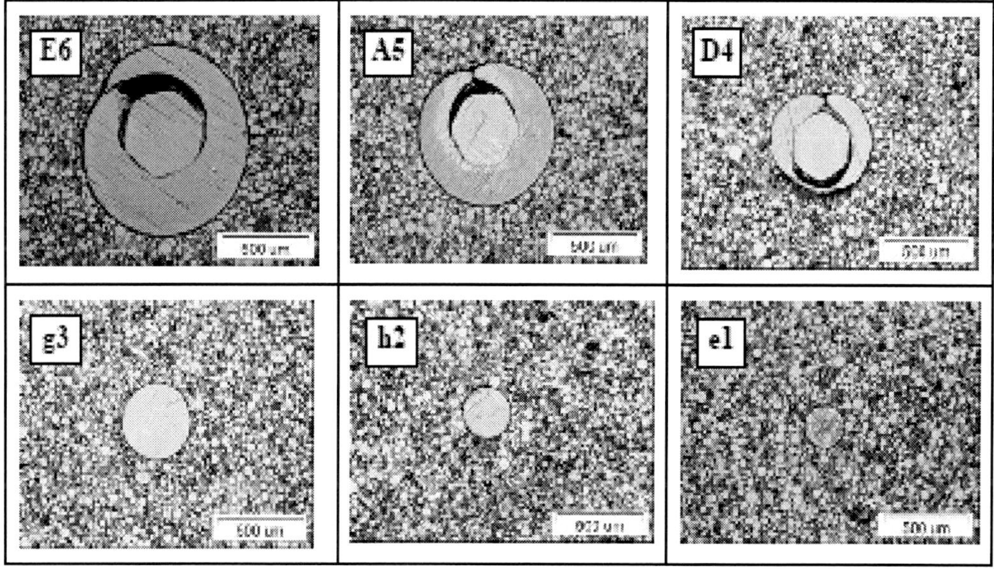

Figure 1. OM of the cross-section of the electric guitar strings marked as E6, A5, D4, g3, h2 and e1 (M=100).

Chemical compositions of the investigated strings are shown in Table 1.

Table 1. Chemical composition of electric guitar strings

Sample		Chemical compositions (%)						
		Mn	Si	Cu	Sn	Ni	Cr	Fe
E6	core	0.502	0.432	0.023	0.333	0.018	0.021	rest
	wrap	0.334	0.212	0.029	0.010	11.10	0.031	rest
A5	core	0.573	0.355	0.011	0.313	0.018	0.003	rest
	wrap	0.356	0.212	0.009	0.012	8.491	0.047	rest
D4	core	0.532	0.343	0.010	0.297	0.002	0.019	rest
	wrap	0.324	0.154	0.006	0.016	8.397	0.001	rest
g3	core	0.570	0.339	0.000	0.302	0.001	0.001	rest
h2	core	0.506	0.293	0.013	0.269	0.000	0.033	rest
e1	core	0.502	0.432	0.023	0.333	0.018	0.021	rest

Comparison of results obtained by chemical analysis (Table 1) as well as results of elemental distribution performed by means of scanning electron microscopy equipped with an X-radiation detector EDS, it can be concluded that investigated E6, A5 and D4 strings are consisting of two parts: steel core wire with Sn coating (Fig. 2A1, 2A2, 2B1, 2B2, 3A) which is hexagonal in the cross section and the steel wrap with Ni coating (Fig. 2A1, 2A2, 2C1, 2C2, 3B) which is round in the cross section. Strings g3, h2 and e1 are consisting of only one steel core with Sn coating (Fig. 2B1, 2B2, 3A) which is round in it's cross section.

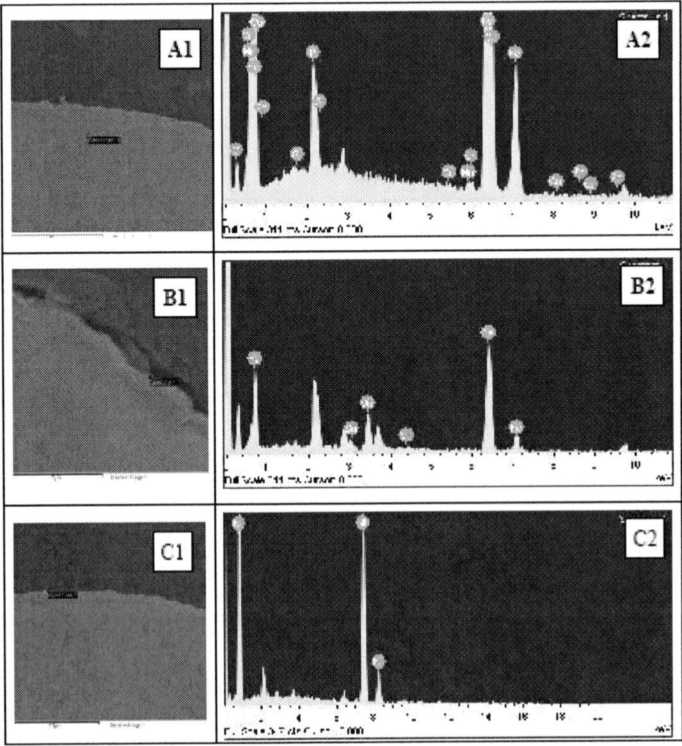

Figure 2. (A2, B2, C2) SEM-EDX spectra results of (A1) core and wrap bulk composition (B1) composition of core coating and (C1) composition of wrap coating, respectively.

On polished cross-sections of E6, A5, D4, g3, h2 and e1 strings Vickers microhardness measurements HV0.1 was performed using indentation load of 0.9807 N for 15 s. Results of Vickers micro hardness data is presented in Table 2.

Table 2. Hardness value of electric guitar string

Hardness number	Sample								
	E6		A5		D4		g3	h2	e1
	core	wrap	core	wrap	core	wrap	core	core	core
HV0.1	633	200	625	200	627	216	625	690	763

Obtained results (Table 2) show that core samples have higher values of Vickers microhardness than wrap samples, which indicates that core and wrap samples have different microstructure.

After chemically etching by immersion polished samples in Nital (2 wt % HNO_3 in ethanol) microstructure of strings were analyzed by optical microscope. Microstructures of core and wrap under optical microscope are showed in Fig. 3. The microstructure of all wrap samples includes typical ferrite structure of low carbon steel (Figure 2A1, 2A2, 3B) and Ni coating. Results of microstructure of all core samples (Fig. 2B1, 2B2, 3A) and their chemical composition (Table 1) indicate that cores of E6, A5, D4, g3, h2 and e1 are made of martenzite steel with Sn layer.

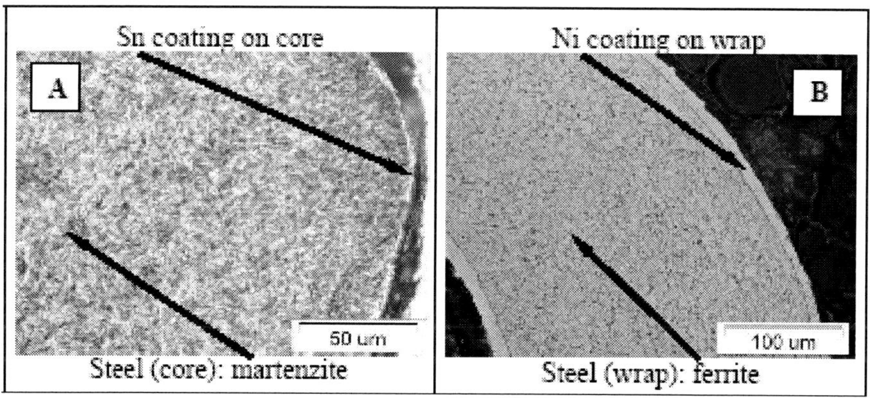

Figure 3. (A) Optical micrograph of cross-section of core of electric guitar string microstructure and (B) wrap of electric guitar string, etched in Nital.

Results show that Vickers microhardness (Table 2) of all cores with martenzite microstructure is higher (HV0.1 values are between 625 and 763). Vickers microhardness for wrap with ferrite microstructure is 200 for E6 and A5 strings and 216 for D4 string.

3.2. Monitoring of Metal Ions Dissolution

The determination of the amount of Ni^{2+}, Sn^{2+}, Si^{4+}, Mn^{2+} and Fe^{3+} ions released in the corrosive solutions after one week of immersion was carried out by means of ICP-OES.

Obtained results presented in Table 3 are expressed as the amount of eluted ions (M^{n+}) in μg per square centimeter and per week of test electric guitar string area (μg M^{n+}/cm^2 week).

Table 3. The amount of eluted ions (Ni^{2+}, Sn^{2+}, Si^{4+}, Mn^{2+} and Fe^{3+}) in μg per square centimeter and per week of test electric guitar string

Sting	μg (Ni^{2+}, Sn^{2+}, Si^{4+}, Mn^{2+} and Fe^{3+})/ cm^2 week				
	Ni^{2+}	Sn^{2+}	Si^{4+}	Mn^{2+}	Fe^{3+}
E6	0.0445	0.8747	0.0578	0.3564	34.9349
A5	0.0196	0.3799	0.0427	0.2195	12.9210
D4	0.0569	0.6940	0.0661	0.2338	10.1118
g3	0.0000	0.3367	0.0800	0.3911	4.3745
h2	0.0002	0.3857	0.0854	0.3761	5.5671
e1	0.0000	0.3462	0.1353	0.8458	5.1913

The largest amount of measured ions in eluates belongs to Fe^{3+} ions. From the results presented in Table 3, it is clear that the amount of dissolved metal ions in corrosive solution is decreasing in the following order: $Fe^{3+}>Sn^{2+}>Mn^{2+}>Si^{4+}>Ni^{2+}$. The lowest amount was noticed for Ni^{2+} ions. The highest amount of eluted Ni^{2+} ions after one weak was 0.0445 μg/cm^2 week from string E6. Those results indicate that the investigated electric guitar strings should not induce contact dermatitis according to the threshold of nickel release, which are below the limits of 0.5 μg cm-^{2} week^{-1}.

Generally, the corrosion of electric guitar strings in artificial sweat solution is determined by the solubility of coating on core and wrap and the solubility of bulk steel wires. After one week of monitoring, the quantities of nickel eluted ions from all investigated strings did not exceed the limits prescribed by standard regulations.

In order to have better insight into the corrosion processes of guitar strings, the dissolution degrees of all eluted metal ions can be calculated. Those results are presented in the following part of this chapter.

A thorough insight into the solubility of individual component of samples can be obtained from the degree of dissolution for each chemical element. This parameter was calculated from the amount of eluted Ni^{2+}, Sn^{2+}, Si^{4+}, Mn^{2+} and Fe^{3+} ions using the following expression:

$$\chi_i = \frac{A}{B}$$

where A and B are respectively the amounts of elements released in the corrosive solution (measured by ICP-OES) and the amounts of elements present in the untreated material. Figures 4 show the relationship between the degree of dissolution of each group of ions and the time of immersion in artificial human sweat for all six string samples investigated (E6, A5, D4, g3, h2 and e1).

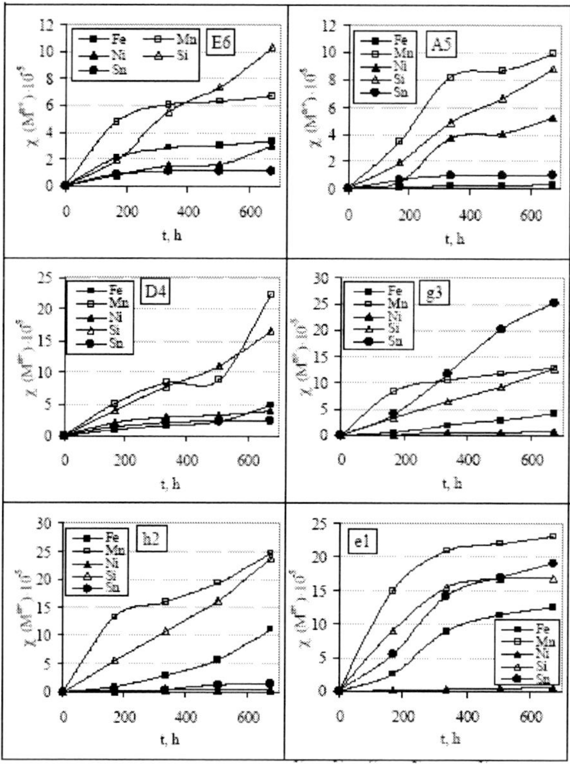

Figure 4. Degree of dissolution (χi) of Ni2+, Sn2+, Si4+, Mn2+ and Fe3+ ions from electric guitar strings versus the immersion time.

The obtained results indicate that during the corrosion of guitar strings in artificial sweat solution some metal ions show significantly higher dissolution degrees than others. Dissolution degrees in majority of samples are decreasing in the following order: χ(Mn) > χ(Si) > χ(Fe) > χ(Ni) > χ(Sn) > χ(Zn). It can be clearly seen that nickel has much lower dissolution degree than iron, which is very good for the protection of skin allergic reaction occurrence. It was to be expected that nickel and tin will have the highest dissolution degrees (since they are coated on the outer surface layers of the strings). On the contrary, these results have shown that other metal ions (manganese, silica and iron) are eluting through the tin and nickel layers, probably by processes of diffusion into the corrosive media.

4. Conclusion

Based on the research results obtained, it has been found that:

1) Investigated E6, A5 and D4 strings are consisting of two parts: steel core wire with Sn coating which is hexagonal in the cross section and the steel wrap with Ni coating which is round in the cross section. Strings g3, h2 and e1 are consisting only of core wire with Sn coating which is round in cross section.

2) According to the obtained results of characterization of investigated strings it is clearly seen that the microstructure of all wrap samples includes typical ferrite crystals of low carbon steel and Ni coating. Results of microstructure of all core samples indicate that cores of strings are made of martenzite steel with Sn layer.
3) Simultaneous determination of the amounts of dissolved Ni^{2+}, Sn^{2+}, Si^{4+}, Mn^{2+} and Fe^{3+} ions from electric guitar strings in the artificial sweat solution can be efficiently monitored by inductively coupled plasma-optical emission spectroscopy (ICP-OES).
4) The amounts of dissolved metal ions in corrosive solution after one week of immersion in the artificial sweat are decreasing in the following order: $Fe^{3+}>Sn^{2+}>Mn^{2+}>Si^{4+}>Ni^{2+}$. Generally, the corrosion of electric guitar strings in artificial sweat is determined by the solubility of coating on core and wrap and the solubility of bulk steel wires.
5) The corrosion of electric guitar strings was monitoring by determination of the degree of dissolution, χ_i of each ion (Ni^{2+}, Sn^{2+}, Si^{4+}, Mn^{2+} and Fe^{3+}) after immersion in artificial sweat. The degree of dissolution in majority of samples are decreasing in the following order: χ(Mn) > χ(Si) > χ(Fe) > χ(Ni) > χ(Sn) > χ(Zn).

REFERENCES

[1] Evans, T; Evans, M.A. *Guitars from the Renaissance to Rock*, Paddington Press Ltd: New York, 1979.

[2] Kraut, RE. *The attraction of the guitar as an instrument of motivation, preference, and choice for use with clients in music therapy: A review of the literature.* The Arts in Psychotherapy 2007, 34, pp 36-52.

[3] Pine, C. *Further Researches in Mathematical Sciences*. Nature 1875, 4, p 6.

[4] Ellis, A. J. *Musical Scales of Various Nations*. Nature 1885, 26, pp 488-490.

[5] Barrett, W. F. *Physical Science for Artists*. Nature 1878, 1, p 356.

[6] Highes, L.E.C. *Electronic Music*. Nature 1940, 145, pp 170-174.

[7] *The Encyclopedia of Musical Instruments,* Zagreb: Znanje, 2007; pp 34-41.

[8] Rezić, I.; Ćurković, L.; Ujević, M. *Metal ion release from electric guitar strings in artificial sweat.* Corrosion Science 2009, 51, pp 1985-1989.

[9] Gambichler, T.; Uzun, A.; Boms, S.; Altmeyer, P.; Altenmüller E. *Skin conditions in instrumental musicians, a self -reported survey.* Contact Dermatitis 2008, 58, pp 217-222.

[10] Önder, M.; Aksakal, A.B.; Öztaş, M.O.; Gürer, M.A. *Skin problems of musicians.* Int. J. Dermat. 1999, 38, pp 192-195.

[11] Baccouche, D.; Mokni, M.; Abdelaziz, A.B.; Osman-Dhahri, A.B. *Dermatological problems of musicians: A prospective stuy in musical students.* Annales de Dermatologie et de Venereologie 2007, 134, pp 445-449.

[12] Gambichler, T.; Boms, S.; Freitag, M. *Contact dermatitis and other skin conditions in instrumental musicians.* BMC Dermatology 2004, 4, pp 3-15.

[13] Smith, V.H.; Holmes, R.C.; Bedlow, A. *Contact dermatitis in guitar players.* Clin. Exp. Dermat. 2005, 31, pp 129-156.

[14] Marshman, G.; Kennedy, *C.T.C. Guitar–string dermatitis*. Contact Dermatitis 1992, 26, p 134.

[15] Hausen, B. M.; Noster, U. *Nickel allergy in amateur guitarists*. Contact Dermatitis 1988, 14, 244-245.

[16] *British Standard* BS EN, 1811, 1999.

[17] Haudrechy, P.; Foussereau, J.; Mantout B.; Baroux, B. *Nickel release from 304 and 316 stainless steels in synthetic sweat. Comparison with nickel and nickel plated metals. Consequences on allergic contact dermatitis.* Corroion Science 1993, 35, pp 329-336.

[18] Fournier, P.G.; Govers, T.R. *Contamination by nickel, copper and zinc during the handling of euro coins.* Contact Derm. 2003, 48, pp 180-188.

[19] Fournier, P.; Grovers, T.R.; Brun A. *Euro coins and the potential risk of nickel allergy.* Europhysics News 2003, 34, pp 1-8.

[20] Nestle, F.O., Speidel, H.; Speidel, M.O. *High nickel release from 1- and 2- euro coins.* Nature 2002, 419, p 132.

[21] Gou, F.; Gleeson, M.; Villette, J.; Kleyn, S.; Kleys, A.W. *The surface of 1-euro coins studied by X-ray photoelectron spectroscopy.* Applied surface Science 2004, 225, pp 47-53.

[22] Vahter, M.; Berglund, M.; Akesson, A. *Metals and women's Health.* Environ Res. 2002, 88, pp 145-155.

[23] Sanchez-Perez, J.; Ruiz-Genao, D.; Garcia Del Rio, I.; Garcia Diez, A. *Taxi driver's occupational allergic contact dermatitis from nickel in euro coins.* Contact Dermatitis 2003,48, pp 337-349.

[24] Perdes Suarez, C.; Fernandez-Redondo, V.; Toribo, J. *Bingo-hall worker's occupational copper contact dermatitis from coins.* Contact Dermatitis 2002, 47, pp 165-166.

[25] Milošev, I.; Kosec, T. *Metal ion release and surface composition of the Cu-18Ni-20Zn nickel-silver durign 30 days immersion in artificial sweat*, Applied surface Science 2007, 254, pp 644-652.

[26] Okazaki, Y.; Gotoh, E. *Comparison of metal release from various metallic biomaterials in vitro.* Biomaterials 2005, 26, pp 11-21.

[27] Liden, C.; Rondell, E.; Skare, L.; Nalbanti, A. *Nickel release from tools on the Swedish market.* Contact Dermatitis 1998, 39, pp 127-131.

[28] Kanerva, L.; Sipiläinen-Malm, T.; Estlander, T.; Zitting, A.; Jolanki R.; Tarvainen, K. *Nickel Release from Metals and a Case of Allergic Contact Dermatits from Stainless Steel.* Contact Dermatitis 1994, 31, pp 299-303.

[29] Milošev, I.; Kosec, T. *Study of Cu-18Ni-20Zn Nickel Silver and other Cu-based alloys in artificial swear and physiological solution.* Electrochimica Acta 2007, 52, pp 6799-6810.

[30] Chiba, A.; Sakakura, S.; Kobayashi, K. *Dissolution amounts of nickel, chromium and iron from SUS 304, 316 and 444 stainless steels in sodium chloride solutions.* Journal of Materials Science 1997, 32, pp 1995-2000.

[31] Colin, S.; Jolibois, H.; Chambaudet, A. Tirefrod, M. *Corrosion stability of nickel in Ni-allyos in synthetic sweat.* International Biodeteriation and Biodegradation 1994, 34, pp 131-141.

[32] Rezić, I.; Zeiner, M.; Steffan, I. *Determination of allergy causing metals from coins*. Monatshefte für Chemie – Chemical Monthly 2009, 140, pp 147-151.

[33] Nölte, J. *ICP Emission Spectrometry*, A Practical Guide. Wiley-VCH: Germany, 2003.

[34] Silva, F.V.; Trevizan, L.C.; Silva, C.S.; Nogueira, A.R.A.; Nóbrega, J.A. *Evaluation of inductively coupled plasma optical emission spectrometers with axially and radially viewed configurations*. Spectrochimica Acta Part B 2002, 57, pp 1905 -1913.

[35] Brenner, I.B.; Zander, A.T. *Axially and radially viewed inductively coupled plasmas- a critical review*. Spectrochimica Acta Part B 2000, 51, pp 1195-1240.

[36] Ivaldi, J.C., Tyson, J. F. *Performance evaluation of an axially viewed horizontal inductively coupled plasma for optical emission spectrometry Spectrochimica* Acta Part B: Atomic Spectroscopy 1995, 50, pp 1207-1226.

[37] Rezić, I.; Steffan, *I. ICP-OES determination of metals present in textile materials*. Microchemical Journal 2007, 85, pp 46-51.

In: Handbook of Research on Chemoinformatics ...
Editor: A.K. Haghi
ISBN: 978-1-62100-998-6

Chapter 4

GREEN CHEMISTRY: A NEED OF THE DAY

Suresh C. Ameta[1], Rakshit Ameta[2], Aarti Ameta[1] and Chetna Ameta[1]

[1]Department of Chemistry, M.L. Sukhadia University, UDAIPUR-313001 (Raj.) India
[2]Department of Pure & Applied Chemistry, University of Kota, KOTA-342005 (Raj.) India

Chemistry plays a very prominent part of our daily lives; of course, sometimes it is undesirable, on one hand, foods and drinks have been made safe to consume, synthesis of new pharmaceutical drugs to cure various dreadful diseases and variety of cosmetics to beautify us, but on other hand, such chemical developments also bring new environmental problems and harmful unexpected side effects. It necessitates the search for harmless or less harmful chemical products which are greener in nature.

In this context, one of the well known example is the pesticide DDT, which was effective in controlling insects pests but it was later on found to have some implications on the bald eagle population and was also suspected to be carcinogenic in nature. Refrigerent like CFCs is quite useful, but it depletes the ozone layer to a great extent, which protects our earth from harmful UV rays of sun.

Such events are considered the genesis of Green Chemistry in the middle of 20th century. The term 'Green Chemistry' was coined by Professor Paul Anastas, who is now also known as the father of green chemistry. Green chemistry involves the development of newer chemical products and synthetic procedures, which are environmentally friendly and have reduced health risks. This environmentally friendly approach to chemistry has been growing rapidly as the threat to our environment and natural resources is become more and more alarming to the scientific community in particular and public in general. The focus of green chemistry focus is to either reduce, recycle or eliminate the use of different toxic chemicals. Much more research is being done these days to find new chemical reactants, processes and ways to eliminate wasteful by-products, while keeping an economical in mind approach. Green chemistry is mainly governed by the twelve basic principles to help us in assessing whether a chemical reaction or a process is 'green' or not? Thus, on applying these basic principles of green chemistry to common conventional organic synthesis, one can either convert or replace it in an efficient, economic and less hazardous green synthesis.

Nowadays green chemistry is becoming one of the most important areas of current research. It has also been referred by a number of synonymous words i.e., Clean Chemistry, Atom Economy, Benign by Design Chemistry, Sustainable Chemistry, Ecofriendly Chemistry, Environmentally Benign Chemistry, etc.

PRINCIPLES OF GREEN CHEMISTRY

Green chemistry is related to the environmentally benign chemical synthesis. It means, it deals with those pathways, which minimize the environmental pollution. T. Anastas [1] gave the following twelve basic principles of green chemistry.

1. Prevention of Waste

This is based on the concept of 'prevention is better than cure', which implies that it is better to carry out a synthesis in such a way so that the formation of wastes (by-products) is either minimum or totally absent, whenever any waste is discharged in the atmosphere, it will not only cause pollution but also requires expenditure for cleaning-up.

2. Maximum Incorporation of all Materials used in the Process into Final Product

A synthesis is considered to be green by nature, when there is maximum incorporation of all the raw materials used in that process into the final product and wasteful by-products are negligible. In this respect, per cent atom utilization is a useful term, which can be determined as –

$$\% \, of \, atom \, utilization = \frac{M.W.of \, desired \, product}{M.W.of \, desired \, product + M.W.of \, waste \, product} X100$$

Similarly, atom economy is as defined by Trost [2] as how much of the reactants ends up in the final product?

$$\% \, of \, atom \, economy = \frac{FW \, of \, atoms \, utilized}{FW \, of \, reactants \, used \, in \, the \, reaction} X \, 100$$

Let us take an example to understand the term atom economy.
Addition of hydrogen to olefin is an atom economic reaction.

$$\underset{\text{Propene}}{H_3C\text{-}CH{=}CH_2} + H_2 \xrightarrow{Ni} \underset{\text{Propane}}{H_3C\text{-}CH_2\text{-}CH_3}$$

In this reaction, all reactants are incorporated into final product. Thus, this reaction is considered as 100% atom economical.

Similarly, all addition, cycloaddition and rearrangement reactions are 100% atom economical, while substitution and elimination reactions are less than 100% atom economical as they either eliminate some molecule or substitute one group by another.

Substitution reaction of ethyl propionate with methyl amine is one of such example.

$CH_3 - CH_2 - C - O - C_2H_5 + H_3C - NH_2 \longrightarrow CH_3 - C - NHCH_3 + CH_3CH_2OH$

|| ||

O O

Ethyl propionate Methyl amine N-methyl Ethyl alcohol propionamide

In this reaction, ethyl alcohol is by-product so the per cent atom economy of this reaction about 65.4%.

3. Prevention of Minimization of Hazardous Products

One of the most important principles of green chemistry is to develop such synthetic methodologies, which generate substances that possess little or no toxicity to the environment and human health.

4. Designing Safer Chemicals

Chemical routes should be designed in such a way that the products like dyes, paints, adhesives, pharmaceuticals etc. so formed are not only more efficacious but also less toxic for a long as well as short term.

5. Safer Solvents and Auxiliaries

A number of aromatic hydrocarbons and halogenated solvents are identified as human carcinogens. Thus, it must be taken into account that the solvent selected for a particular reaction should not cause any environmental pollution or health hazard. If possible, the use of solvent should be avoided or the reaction is carried out on the solid support. If necessary at all, they should be used as little or mildly as possible, so that their effects is minimized.

6. Design for Energy Efficiency

The chemical synthesis should be energy efficient it means that energy consumption during the reaction should be kept to a minimum. Process like distillation, recrystallization, refluxing, ultrafiltration etc. requires a large amount of energy. Thus, such process should be designed so that there is no need of separation and purification. It this way, one can minimize the energy consumption to a large extent. Further, this energy can also be supplied to a reaction by light, microwave, ultrasound, etc.

7. Use of Renewable Feedstock

Starting materials can be obtained either from renewable or non-renewable materials. In that case, raw materials should be renewable as far as possible because non-renewable materials are very essential and can't be regenerated again. Materials obtained either naturally (CO_2, methane gas) or from agriculture and biological processes are normally categorized as renewable starting material.

8. Reduction of Derivatization

Derivatization, means protecting or blocking groups during the chemical synthesis of the desired chemicals, pharmaceuticals, pesticides, and certain dyes, etc. A typical example is given here to understood derivatization.

NH_2 / C≡N $\xrightarrow{C_6H_5CH_2Cl}$ $NHCH_2C_6H_5$ / C≡N $\xrightarrow[SnCl_2]{HCl}$ $NHCH_2C_6H_5$ / CHO $\xrightarrow{\text{Deprotection}}$ NH_2 / CHO

m-aminobenzonitrile m- aminobenzaldehyde

In this synthesis, amino group is to be protected by using benzyl chloride (a known hazardous material) in order to carry out the reaction at other group (-C≡N) present in molecule. But after deprotection benzyl chloride again regenerate as a waste. Thus, making this reaction less atom economical and therefore, derivatization to be is avoided if possible to make any process green chemical.

9. Use of Catalyst

The catalyst facilitates the transformation of reactants into products without being consumed. Sope of more advantages of catalyst are listed below:

(i) Better yield

Toluene (CH_3) —[O], Aq. $KMnO_4$, Crown ether→ Benzoic acid (COOH) [85% yield]

(ii) Those reactions which are not feasible, can occur in the presence of catalyst; e.g.

$HgSO_4$

$CH \equiv CH + H_2O \longrightarrow H_3C - CHO$

H_2SO_4

(iii) Selective enhancement

CH_3

NaOEt |

$C_6H_5 - C - CH_2COCH_3 \longrightarrow C_6H_5 - C - CH - COCH_3$

CH_3I ||

O O

Here, C-methylation is preferred over O-methylation.

10. Designed Product should be Biodegradable

The chemical synthesis should be designed in such a way that products so formed must be biodegradable. If these are not biodegraded, then they remain in the environment for a long and during the course of time, get accumulated in various plants and animal species and later on incorporated in their food chain causing harm to the concerned species. The problem is encountered especially for pesticides and polymers. Thus, one should synthesize such molecules, which may posses functional group that facilitates its degradation either through hydrolysis, photolysis or other methods of cleavages.

11. Designing Safer Manufacturing Plant

The gas tragedy in Bhopal and Seveso (Italy) and many other such chemical accidents have resulted not only in loss of a large number of lives but they have also made many persons disabled for the rest of their lives. Thus, manufacturing unit should be designed in such a way that any possibilities of lethal accidents are avoided including storage and handling of toxic chemicals.

12. Strengthening of Analytical Techniques

Newer analytical technologies should be developed, which required minimum use of chemicals, so as to prevent generation of waste hazardous materials. Further, accurate and reliable sensors must be designed and used to monitor the hazards generated during the process to prevent any possible accident in chemical plant.

SOME GREEN CHEMICAL REACTIONS

With the advancement of science and development of newer and better techniques, a large number of organic reactions could be carried out in ecofriendly conditions. It has now been possible to carry out the reaction in aqueous phase and in solid state i.e., without solvent. In number of cases, such reactions occur more efficiently and with more selectivity, compared to reactions carried out in solvents. Such reactions are now designated as Green Reactions.

1. Aqueous Phase Reactions

As we all know that water is considered as a universal solvent. It has unique properties of enthalpy, entropy, low volatility and no carcinogenic nature, which has led the chemists to use quite commonly it as a solvent. Some reactions, which have been carried out in aqueous medium are:

a) Aldol condensation:

Self condensation of aldehydes (having α-hydrogen atom) in the presence of dilute alkali to give β-hydroxyaldehdyes (known as aldols) is called aldol condensation.

OH

^{-}OH |

$CH_3CHO + CH_3CHO \longrightarrow CH_3 - CH - CH_2CHO$

Acetaldehyde β-hydroxy butyraldehyde

Earlier aldol condensation of silyl enol ethers was carried out in anhydrous solvent[3,4] but now it has become possible to perform such aldol condensation of silyl enol ethers with aldehdyes in aqueous phase.[5,6]

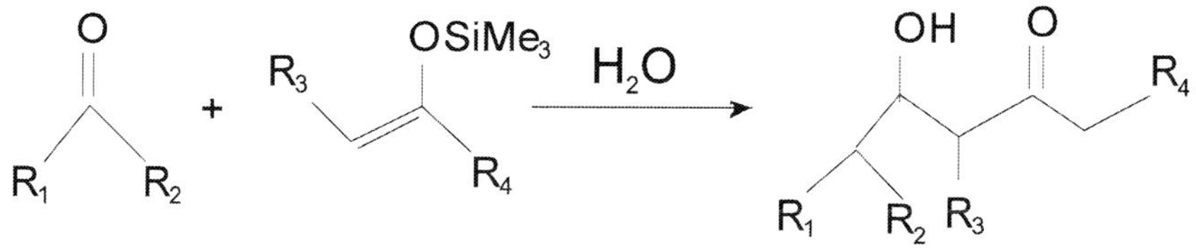

This condensation reaction has been carried out without any catalyst, but it takes several days for completion, since water acts as a week Lewis acid. Therefore, addition of stronger Lewis acid like ytterbium trifate increases the yield as well as the rate of this reaction [7].

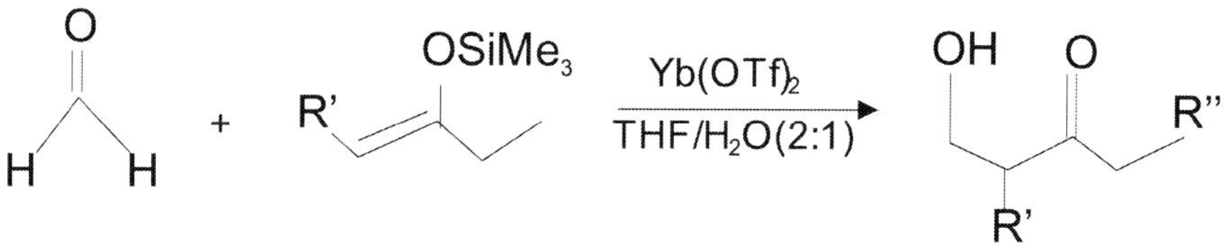

(97-98%)

b) Diels-Alder reaction

It is a [4 + 2] cycloaddition reaction between a conjugated diene (4π electron system) and a dienophile the compound having a double bond (2π electron) to form a cyclic structure [8,9].

Diene Dienophile Cyclic adduct

Diels-Alder reaction proceeds satisfactorily in aqueous phase [10,11]. Reaction of furan with maleic acid in hot water gives an adduct.

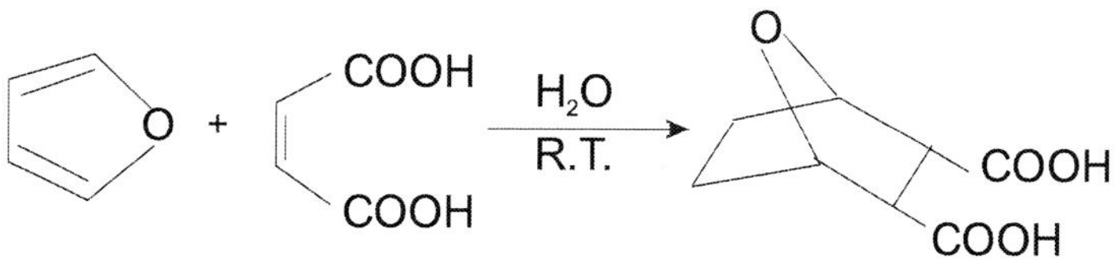

Similarly

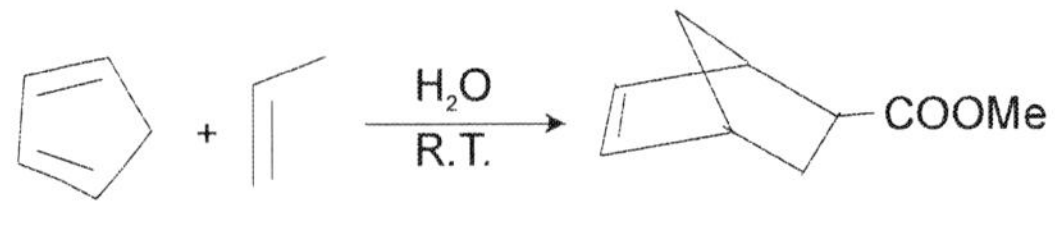

Cyclopentadiene Methyl Endo + Exo. Adduts

acrylate

The intramolecular azo-Diels-Alder reaction [12] also takes place in aqueous media. This reaction gives fused ring system as adduct with reasonably good yield.

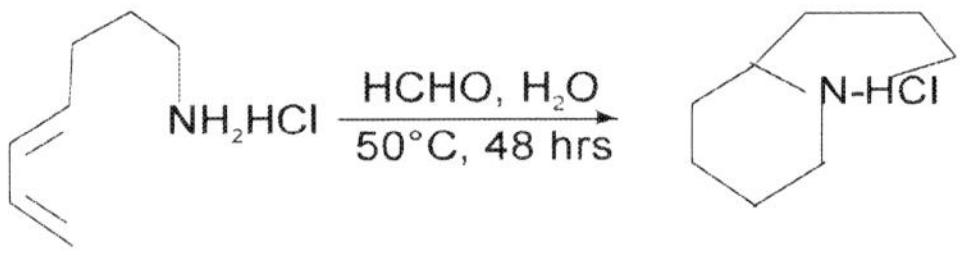

(95%)

c) Knoevenagel condensation

Base catalyzed condensation between an aldehyde or ketone, with compound having active methylene group (malonic ester) is called Knoevengel condensation [13].

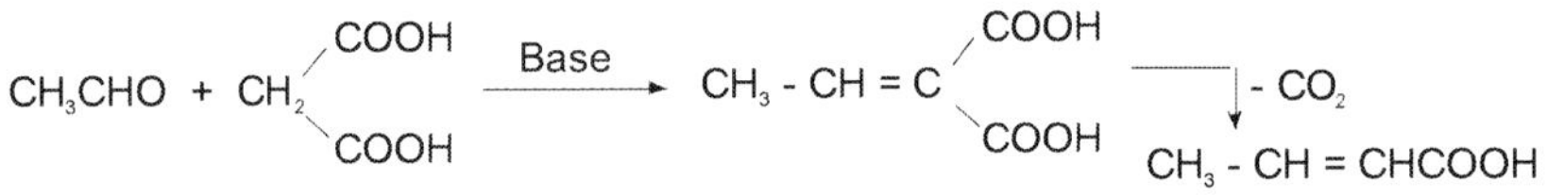

Acetaldehyde Malonic acid Crotonic acid

The Knoevenagel reaction has ALSO been reported between adehydes and acetonitrile in water. Salicyaldehydes react with malononitrile in heterogeneous aqueous alkaline medium to give o-hydroxy benzylidenemalononitrile [14,15] which then converted into coymarins with good yield on acidification. Recently this condensation is revisited [16] and synthesized new dimeric chromene derivatives.

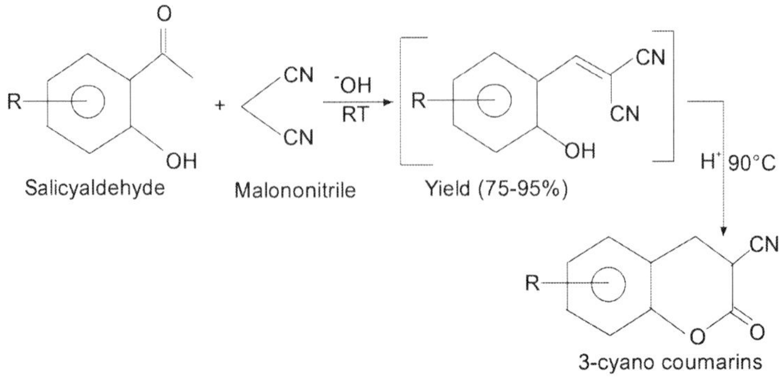

The condensation of substituted acetonitriles with salicyaldehydes requires the presence of CTAB. It was found that aqueous phase reaction gives better yield.

R	Yield of coumarins (%)	
	In water	In ethanol
CN	90	70
COOEt	66	35
2-Py	98	55
NO_2	87	80

d) Benzoin condensation

Aromatic aldehydes (having no α-hydrogen) on treatment with sodium or potassium cyanide undergo self condensation in presence of aqueous ethanolic solution giving α-hydroxyketone (benzoin) [17,18]. This reaction is known as benzoin condensation.

H   O OH

| CN^-, Δ || |

2 Ph – C = O ⟶ Ph – C – C – Ph

EtOH, H_2O |

H

Benzaldehyde   Benzoin

Further, Kool and Breslow [19] osberved that the benzoin condensation in aqueous media using inorganic salts (LiCl) in about 200 times faster than in ethanol (without any salt) and Iwamoto [20] reported reaction in the presence of benzimidazolium salt.

e) Oxidation

One of the most widely investigated process in organic chemistry is oxidation. The oxidation of arenes with $KMnO_4$ in aqueous alkaline medium is quite known.[21]

(i) Oxidation of alkenes: The oxidation of alkenes with aqueous solution of $KMnO_4$ in presence of phase transfer catalyst $[CH_3(CH_2)_{15}N^+ (CH_3)_3Cl^-]$ or crown ether gives corresponding carboxylic acid.

aq. $KMnO_4$

$CH_3(CH_2)_5CH{=}CH_2$ ——→ $CH_3(CH_2)_4CH_2COOH$

PTC

1-Octene Heptanoic acid

The oxidation of n-octane to 1-octanol using *Pseudomonas oleovorans* gives 98% yield.

Pseudomonas oleovorans

$CH_3(CH_2)_6CH_3$ ——→ $CH_3(CH_2)_6CH_2OH$

H_2O

1-Octanol (98% pure)

(ii) Oxidation of alkynes: Alkynes are oxidized with $KMnO_4$ in aqueous medium to give a mixture of carboxylic acids.

$KMnO_4$

$R - C \equiv C - R' + 4\,[\,O\,] \longrightarrow RCOOH + R'COOH$

Aq.

(iii) Oxidation of aldehydes and ketones: Aromatic aldehydes have been conventionally oxidized by aqueous performic acid (H_2O_2 + HCOOH) at low temperature. [22]

2. Solid Phase Organic Synthesis

Reactions carried out in solid phase [23-27] i.e., either without using solvent or using any solid support, are easy to handle, reduce pollution, low cost to operate. Such reactions are especially important in industries. It is believed that solvent free organic syntheses are industrially useful and largely green in nature.

a) Aldol condensation

The aldol condensation of lithium enolate [28] of methyl 3,3-dimethylbutanoate with aromatic aldehydes gives a mixture of Syn- and Anti- products in 8 : 92 ratio.

RCHO + (Lithium enolate: t-Bu, OLi, OMe) —Solid, 3 days / RT, Vaccum→ Syn-product + Anti-product

Aromatic aldehyde

Lithium enolate

OH O, R, t-Bu, OMe + OH O, R, t-Bu, OMe

Where, R = 2-$OCH_3C_6H_4$– Syn-product 8 : 92 Anti-product

4-$NO_2C_6H_4$ -, 2-$NO_2C_6H_4$ –

4-NO_2-2-thenyl (Yield 70%)

In absence of any solvent, some crossed aldol condensation [29] proceed more efficiently and that too with stereoselectivity. In this procedure, aldehyde and ketone and NaOH are grounded in pestle and mortar at room temperature for 5 min and chalcone was obtained.

ArCHO + ArCOMe —NaOH / Solid State→ [Ar CH(OH)CH_2 COAr'] (Aldol) —H_2O→ Ar–CH=CH–COAr' (Chalcone)

Ar	Ar'	Reaction time (min.)	Chalcone yield (%)
p-Me C_6H_4	Ph Ph	5	97
p-Me C_6H_4-	p-MeC_6H_4-	2	99
p-Cl C_6H_4 -	Ph	5	98
p-Cl C_6H_4	p-MeOC_6H_4 -	10	79

b) Pinacol-pinacolone rearrangement

Pinacol-pinacolone rearrangements proceed faster and more selectively in solid state. [30]

Ph–C(H)(OH)–C(Ph)(OH)–R —H^+→ Ph–C(=O)–C(Ph)(H)–R + H–C(=O)–C(Ph)(Ph)–R

Pinacol

c) Benzil-benzilic acid rearrangement

Normally this reaction is carried out by heating benzil and alkali metal hydroxides in presence of aqueous organic solvent, but it was found that the rearrangement proceeds more efficiently and faster in solid state. [31]

Ar'

KOH

Ar – C – C – Ar' ⟶ Ar — COOH

|| || Solid

O O   0.1 – 6 hrs   OH

Benzil   Benzilic acid (70-93% Yield)

Ar = Ar' = Ph, p-ClC_6H_4, p-$NO_2C_6H_4$

Ar = -Ph   Ar' = p – ClC_6H_4-

Ar = -Ph   Ar' = p-$NO_2C_6H_4$-

d) Beckman Rearrangement

Solid state Beckman rearrangement [32] has been carried out with oxime of ketone loaded on montmorillonite and irradiated for 7 min in a microwave oven to give corresponding anilide in high yield (91%) While conventional heating gives only 17% yield.

CH_3

C = NOH Mont. K10 / Microwave / 7-10 min ⟶ Me - C(=O) - N–Ph

Acetanilide (91%)

e) Nitration

The nitration of aromatic compounds in the presence of nitric acid and acetic anhydride at 0-20°C with zeolite beta catalyst (in absence of any solvent) gives all the tree isomers.

R

HNO_3/Ac_2O / Zeolite beta ⟶

R, NO_2 + R, NO_2 + R, NO_2

o-Nitro   m-Nitro   p-Nitro

R	Yield (%)	Ratio of isomers (%)		
		o-	m-	p-
Me	>99	18	3	79
t-Bu	92	8	Traces	92
Cl	>99	7	0	33
NO_2	13	1	92	7

SOME OTHER GREEN PATHWAYS

There are some other pathways also, from which one can design green synthesis. These are:

1) Microwave induced green synthesis

In microwave induced organic reactions, the reactions are carried out in a solvent (medium) or on a solid support, where no solvent is used. For reaction in a solvent, the choice of the solvent becomes very important. [33-35] A commonly used solvent in a domestic microwave oven is N,N-dimethylformamide (DM). This solvent can retain water formed in a reaction and thus, obviates the need to separate water. A major advantage of using microwave as a green chemical pathway is substantial savings of time and secondly, the reactions can be carried out in water or more commonly under solvent free conditions (solid state). A number of organic synthesis have been carried out in presence of microwaves and these syntheses are grouped mainly under three categories:

(i) Microwave – assisted reactions in water
(ii) Microwave – assisted reactions in organic solvents
(iii) Microwave solvent – free reactions (solid state reactions)

(i) Microwave-assistant reactions in water

a) Hydrolysis: Microwave radiations have been extensively used for hydrolysis reactions.

- Hydrolysis of benzyl chloride

Hydrolysis of benzyl chloride with water in microwave oven gives 97% yield [36] of benzyl with alcohol in 3 min while conventionally it takes about 35 min.

$$C_6H_5CH_2Cl + H_2O \xrightarrow[3\ min]{MW} C_6H_5CH_2OH$$

Benzyl chloride Benzyl alcohol (97%)

- Hydrolysis of benzamide

Under microwave conditions, the acid hydrolysis of benzamide is completed in 7 min. giving 99% yield of benzoic acid. [37]

20% H_2SO_4

$C_6H_5CONH_2 \longrightarrow C_6H_5COOH$

MW 7 min

Benzoic acid (99%)

- Hydrolysis of N-phenyl benzamide

Hyrdolysis of N-phenylbenzamide takes 12 min. under microwave exposure.

20% H_2SO_4

$C_6H_5CONHC_6H_5 \longrightarrow C_6H_5COOH$

MW 12 min

Benzoic acid (74%)

- Hydrolysis of methylbenzoate to benzoic acid (Saponification)

Saponification of methylbenzoate in aqueous sodium hydroxide under microwave conditions gives 84% benzoic acid. [38,39]

aq. NaOH

$C_6H_5COCH_3 \longrightarrow C_6H_5COOH$

MW 2.5 min

Benzoic acid (84%)

b) Hofmann elimination

Hofmann elimination in presence of microwave irradiations gives high yield, normally unstable product water-chloroform system is used.

O
$\overset{+}{N}R_3I^-$
OEt

Water/$CHCl_3$
MW; 1 min

O
OEt

c) Oxidation of alcohols

Primary alcohols are oxidized to the corresponding carboxylic acid using sodium tungstate in 30% aq. hydrogen peroxide.

R OH

30% H_2O
NO_2WO_4 ; MW

O
R OH

Primary alcohol Carboxylic acid

Thiols are oxidized to disulphides on mineral supports like silica, celite, florisel or alumina.

$$2R - SH \longrightarrow R - S - S - R$$

R – SH = p-Nitrophenylmercaptan, 2-Mercaptopyridine, Thiosalicylic acid.

(ii) Microwave-assisted reactions in organic solvents

a) Esterification

A mixture of benzoic acid and n-propanol on heating in a microwave oven for 6 min in presence of conc. sulphuric acid gives propylbenzoate. [40]

Conc. H_2SO_4

$$C_6H_5COOH + n\text{-}C_3H_7OH \longrightarrow C_6H_5COOC_3H_7$$

MW; 6 min Propylbenzoate (70%)

Recently enzymatic esterification by fluorescence spectroscopy was also reported. [41]

Carboxylic acid and benzyl ether on heating in a microwave oven for 2 min gives esters in presence of $LnBr_3$.[42]

$LnBR_3$

$ArCH_2OR + R'COOH \longrightarrow ArCH_2OCOR' + ROH$

Ln = Ld, Nd, Sm, Dy, Er

MW; 2 min

b) Decarboxylation

Conventionally yields are very low for decarboxylation reactions but in a microwave decarboxylation takes place in a much shorter time with high yields under exposure. [43]

MW; 12 min

Quinolone

6-Methoxyindole-2-carboxylic acid → 6-Methoxyindole (99%)

c) Synthesis of chalcone

Microwave radiations have been used for chalcones synthesis. [44] An enhancement in rate was observed, when this reaction performed in presence of microwave radiations. The reaction time was also lowered down from hours to min and that too with improved yields.

Ketone + OHC-Ar → Chalcones (90-100%)

EtOH

NaOH

MW; 30 sec - 2 min

d) Diels-Alder reaction

Conventionally this reaction requires 90 min reflux. However, under microwave conditions [45] high yield of the adduct was obtained in 90 sec using deglyme as a solvent.

Diglyme

MW; 90 sec

Adduct (80%)

e) Fries rearrangement

This rearrangement is useful for the preparation of phenolic ketones. There is considerable rate enhancement of Fries rearrangement carried out in microwave ovens as compared to conventional methods. [46]

$$\text{p-Cresyl acetate } (p\text{-}CH_3C_6H_4OCOCH_3) + AlCl_3 \xrightarrow[\text{MW; 2 min}]{\text{Chlorobenzene}} \text{2-Hydroxy-4-methyl acetophenone (85\%)}$$

(iii) Microwave solvent free reactions (solid state reactions)

In this, the reactants are dissolved in a suitable solvent like water, alcohol, methylene chloride etc. and then the solution was stirred with a suitable adsorbent or solid support like silica gel, alumina and zeolite [47-49]. After stirring, the solvent was removed in vacuo and the dried solid support, on which the reactants have been adsorbed, are used for carrying out the reaction under microwave irradiation. Following are some of the examples of solid state synthesis.

a) Synthesis of heterocyclic compounds

Heterocyclic compounds are very important due to their wide spectrum of biological activities. A number of heterocyclic systems can be synthesized in high yields by using microwave irradiation [50-52].

b) Reduction

Sodium borohydride has been extensively used as a reducing agent. A microwave assisted reaction has been developed for the reduction of aldehydes and ketones using alumina supported $NaBH_4$.

$$R\text{-}C_6H_4\text{-}C(=O)\text{-}R_1 \xrightarrow[\text{MW}]{NaBH_4\text{ - Alumina}} R\text{-}C_6H_4\text{-}CH(OH)\text{-}R_1$$

c) Synthesis of anhydrides from dicarboxylic acids

Anhydrides can be prepared from dicarboxylic acids in presence of isopropenyl acid (which acts as water scavenger) under microwave irradiations using montmorillonite – KSF. [53]

$$(CH_2)_n(COOH)_2 \xrightarrow[MW]{\text{Montmorillonite KSF}} (CH_2)_n(CO)_2O$$

d) Deprotection

The carboxylic group is protected by the benzyl group. After the reaction, the deprotection is carried out by using different reagents like K_2SO_3, $AlCl_3$, $Na\text{-}NH_3$ etc. This process requires a longer time and gives moderate yields but the microwave irradiation procedure [54] is completed within 3-10 min. and yield is also quite high 89-92%.

$$R\text{-}C_6H_4\text{-}COOCH_2C_6H_5 \xrightarrow[\text{MW; 7 min}]{\text{Acidic alumina}} R\text{-}C_6H_4\text{-}COOH$$

R = H or CH_3 (89 - 92%)

e) Synthesis of nitriles from aldehydes

This conversion is achieved by converting aldehyde into oxime followed by its dehydration [55] but this method has a number of drawbacks. In microwave radiations, the aldehyde is converted into oxime by reaction with hydroxylamine hydrochloride and potassium fluoride on alumina (solid support). This oxime is then converted into nitrile by treating it with carbon disulphide. This method gives nitriles in good yield. [56]

$$R-CHO \xrightarrow[\text{(ii) } CS_2,\ 20^{\circ}C]{\text{(i) } NH_2OH\ HCl \text{ on } Al_2O_3 - KF\ MW;\ 5\ min} RC \equiv N$$

2) Ultrasound assisted green synthesis

Ultrasound refers to sound waves having frequencies higher than available frequencies, to which the human ear can respond (ν = 16 KHz). The ultrasound has also been used for chemical synthesis. [57-59] The ultrasound is generated with the help of an instrument having ultrawave transducer, which convert electrical energy into sound energy. The term sonochemistry is used to describe the effects of ultrasound waves on chemical reactivity. Ultrasound assisted organic synthesis also gives good yields. Ultrasound is helpful in number of organic transformation and it also stimulate some microbiological reactions. The symbol))) is used for ultrasound waves exposure.

(i) Saponification

Conventionally saponification of the ester with aqueous alkali required 20 min and the product obtained is in low yields (15%). However sonication [60] gives 94% yield in 60 min.

COOCH3 CH3 ⁻OH / H2O 60 min → COOH CH3

CH3 Methyl 2,4-dimethyl benzoate → CH3 2,4-Dimethyl benzoic acid (94%)

(ii) Hydrolysis

Nitriles can be hydrolyzed to carboxylic acids under basic condition upon sonication. [61]

$^{-}OH / H_2O$

ArCN ——————→ ArCOOH

Ester hydrolysis under sonication also reported. [62]

(iii) Reduction

Zn-NiCl2 (9:1) / EtOH-H2O (1:1)
RT, 2.5 hr
(97%)

Zn-NiCl2 (9:1) / EtOH-H2O (1:1)
RT, 3 hr
(96%)

Zn-NiCl2 (9:1) / EtOH-H2O (1:1)
pH 8.0 RT, 1.5 hr

Sonication increases the activation of nickel powder [63] which is used for the reduction of alkenes. The mixture of nickel chloride and zinc powder gives active nickel. This is used as a catalyst for the reduction of C = C in α,β-unsaturated carbonyl compounds. In absence of ultrasound reaction completed in 48 hrs while in presence of ultrasound, it is over in less than 3 hrs.

(iv) Coupling reactions

Coupling reactions of aryl/alkyl with lithium in THF takes place on sonication [64], while reaction takes place in absence of ultrasound.

Li, THF

$C_6H_5Br \longrightarrow C_6H_5 - C_6H_5$

Similarly, coupling of benzyl halide in presence of copper or nickel powder generated by lithium reduction of corresponding halides in presence of ultrasound gives high yields of dibenzyl. [65]

$C_6H_5Cl + Cu \longrightarrow C_6H_5CH_2CH_2C_6H_5$

LiLi / THF

$CuBr_2$

(v) Diels-Alder reaction

The addition of dimethylacetylene dicarboxylate to furan in water under sonication gives good yield (99-100%) of adduct, while conventionally the yield of this adduct is only 59%.

Y, X, O, COOMe, COOMe, H_2O, 22-45°C, X, Y, O, COOMe, COOMe

(99-100%)

(vi) Cannizaro reaction

The Cannizaro reaction (catalyzed by barium hydroxide) is also accelerated under sonication, whereas there is no reaction without ultrasound. [66]

$Ba(OH)_2$, EtOH

$C_6H_5CHO \longrightarrow C_6H_5CH_2OH + C_6H_5COOH$

10 min (100%)

(vii) Conversion of ketones into tertiary alcohols

Under sonication, ketones on treatment with alkyl halide and lithium in THF gives good yields of tertiary alcohols. [67] This reaction is known as Barbier reaction.

$$R_1R_2C=O \xrightarrow[\text{RT, 10-15 min}]{\text{RX/Li/THF}} R_0R_1R_2C\text{-}OH \quad (70\text{-}100\%)$$

$$\xrightarrow[\text{RT, 15 min}]{\text{BuBr. Li/THF}} \quad (90\%)$$

(viii) Strecker synthesis

The Strecker synthesis of aminonitriles proceeds under sonication. [68]

$$R_2CO \xrightarrow{RNH_2,\ KCN,\ AcOH} R_2C(CN)(NHR')$$

3) Photocatalytic green synthesis: Photochemistry is the branch of chemistry deals with study of physical and chemical changes, which take place in molecules on absorption of suitable radiation ranging from 200-800 nm. The light radiation of the visible and UV regions are responsible for such reactions and these reactions are known as photochemical reactions. [69,70]

A large number of photochemical reactions were studied like nitration of aromatic compound reported by Ameta et al. [71]

hν •

$$NO_2 \longrightarrow NO_2 + OH^-$$

$$C_6H_6 + NO_2 \longrightarrow C_6H_5NO_2$$

As it has been known that the slow kinetics achieved by photochemical reactions can be enhanced by the addition of metallic salts, ozone, semiconductors like ZnO [72], TiO_2 [73,74], CdS [75], etc.

Green Solvents: Most of the chemical reactions have been carried out in organic solvents and these organic solvents are the major source for the emission of volatile organic compounds (VOC).

This emission is reduced by 50% in Germany within the last decade (1990-2000). This reduction was mainly achieved in the transportation and fuel sector but the total emission from solvent used still remains the same.

It shows that emission of VOC can be partly controlled by technical measures but their reduction requires replacement of existing solvents. Thus use of green solvents has a large potential to contribute towards environmentally friendly processes in chemical and pharmaceutical industries. Green solvents are biosolvents derived from the processing of agricultural crops like ethyl lactate which is derived from processing corn.

The major problem of using organic solvents is their adverse environmental impact and hence, organic chemists need some benign process. For this, there are two ways:

a) Solvent free reactions: These conditions include several advantages like –

(i) There is no reaction media to be collected, purified and recycled.
(ii) There is no need for any specialized equipment.
(iii) Chromatographic procedure is not required to separate the product as sufficiently pure compounds are formed.
(iv) Greater selectivity of such reactions.
(v) Less reaction time required for these reactions and that too, with increase in yields.
(vi) These reactions are economically favourable

But there are still some problems associated along with these advantages as lack of solvents may lead to over heating of the reaction mixture and as a result yields by-products.

b) Use of Green solvents: The implementation of green solvents in organic reactions leads to improvements and generation of green solvents. The first generation of green solvents are ordinary organic solvents having reduced adverse ecological impacts as compared to traditional reaction media (solvents), while second generation green solvents consist of water, carbon dioxide, liquid polymers, ionic liquids, and super critical fluids (SCF).

Ethyl lactate: It is ethyl ester of lactic acid. Ethyl lactate and other ester solvents are commonly used solvents in the paints and coating industries because of their high solvency power, high boiling point, low vapour pressure and low surface tension. Side wise, ethyl lactate is 100% biodegradable, non-corrosive, non-carcinogenic, no-ozone-depleting and also easy to recycle. Therefore, it is a desirable as coating for wood, polystyrene and metals and it also acts as a very effective paint stripper. Ethyl lactate replaced other organic solvents like toluene, acetone and xylene. The other applications Ethyl lactate is also used as an excellent cleaner in the polyurethane industries. It is because of its high solvency power as it is able to dissolve a wide range of polyurethane resins. It can also be used to clean a variety of metal surfaces, to remove greases, oils, adhesives and solid fuels. It has almost eliminated the common use of chlorinated solvents.

WATER

One of the most attractive alternatives to volatile organic compounds is water. It can be used as a solvent for many of the organic reactions. It provides one of the best solutions to the existing problem of solvent toxicity and its disposal. It is readily available, plentiful and low cost also. Water is a very useful solvent for synthetic chemistry in terms of the economic and less hazardous production of chemicals.

Sijbren and Engberts [76] reported that one can achieve unusual and unique selectivity in a number of organic reactions by using water as a solvent to favour transition status that optimizes hydrophobic interactions.

Breslow and Biscoe observed that the Dies-Alder reactions in water proceed with relatively higher rates and more endoselectiivty and exoselectivity as compared to reactions carried out in organic solvents. In this reaction the solvent water favours a more compact endotransition state and the accelerating effect is due to hydrophobic effect as well as hydrogen bonding between water and reactants. [77,78] they also showed that hydrophobic interactions might determine the ratio of O-alkylation / C-alkylations of phenoxide ions when water is used as a solvent.

The regio-selective reduction of naturally occurring sulphated steroids was by using hydrophobic borohydrides in water. When reaction was carried out in water using $LiBH_4$, the reduction of these steroids proceed with selectivity for more reactive 17-keto group.

By using hydrophobic oxidizing agent, the epoxidation of olefins could be selectively accelerated [79,80]. Cheng *et al.* [81] studied that water at high temperature can enhance the selectivity and suppress the formation of coke significantly in the catalyzed Fe-Mo based hydrogenation of naphthalene to produce tertalin.

Baruwati *et al.* [82] discovered glutathione promoted green synthesis of silver nanoparticles in water using microwaves. Glutathione is an absolutely benign antioxidant, which acts as a reducing as well as capping agent in aqueous medium.

Carbon dioxide: It is natural, low cost, plentiful and available in 99.9% pure form. It is by-product of brewing, ammonia synthesis or combustion. CO_2 as a solvent has already been used in a variety of commercial processes. It is non-toxic, but acts as a asphyxiant at higher concentrations. CO_2 can be disposed of with no net increase in global CO_2 as it can be easily removed and recycled. Products in CO_2 can be easily isolated by evaporation to 100% dryness. There is no solvent effluent in isolation. CO_2 is the final product of oxidation and hence, it is fully inert under oxidizing conditions. Because of this property CO_2 is used as a special medium for aerobic oxidation.

IONIC LIQUIDS AS GREEN SOLVENTS FOR THE FUTURE

For the last two millennia, most of chemical reactions have been carried out in some or the other organic or molecular solvents. Recently, a new class of solvent has been developed as ionic liquids. Ionic liquids, ionic fluids, fused salts, liquid salts or ionic glasses, are liquids, which are comprised predominantly of some ions and ionic pairs at some given temperature.

Several authors have described the field of ionic liquids. [83-86] The first ionic liquid $[EtNH_3]$ $[NO_3]$ working at room temperature (m.p. 12°C) was discovered in second decade of last century, but the discovery of binary ionic liquids made from mixture of aluminium (III) chloride and n-alkylpyidinium [87] or 1,3-dialkylimidazolium chloride [88] were developed later on, which created interest in this field.

Ionic liquids are divided mainly in two categories: (i) Simple salts which are made of a single anion and cation and (ii) Binary ionic liquids salts, where an equilibrium is involved.

For example $[EtNH_3]$ $[NO_3]$ is an example of simple salt, whereas mixture of aluminium (III) chloride and 1,3-dialkylimidazolium chloride represents an example of binary ionic liquid system containing several different ionic species. Their melting point and other properties depend upon the mole fractions of the two components i.e., aluminium (III) chloride and 1,3-dialkylimidazolium chloride.

$[PF_6]^-$ $[BF_4]^-$ C_6H_{13} $[NO_3]^-$

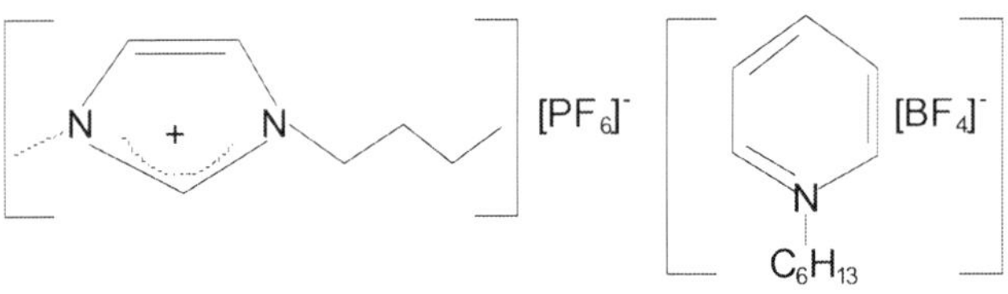

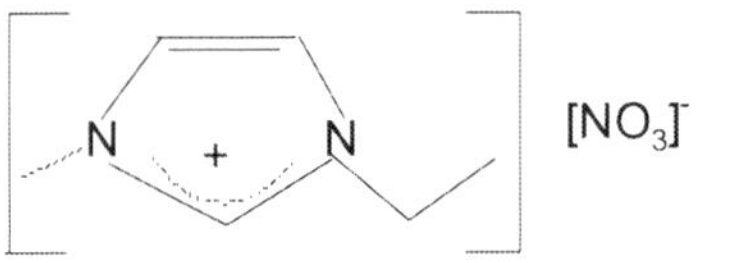

Simple Room Temperature Ionic Liquids

Characteristics: Ionic liquids are having moderate to poor electrical conductivity, non-ionising (non-polar), highly viscous and exhibit a low vapour pressure and hence, can not emit vapours like other volatile organic compounds. They have some diverse other properties like low combustibility, excellent thermal stability and favourable solvating properties for a large range of polar as well as non-polar compounds.

The miscibility of these ionic liquids with water or organic solvents varies with length of side chain on their cation and also on the choice of counter anion. The solubility of different species in imidazolium ionic liquids depends mainly on polarity and hydrogen bonding ability. For example, saturated aliphatic compounds are generally very low solubility (sparingly soluble) in ionic liquids, while olefins show relatively more solubility and compounds like aldehydes are almost completely miscible. Due to such properties, ionic liquids can be used in biphasic catalysis (hydrogenation and hydrocarbonylation), which makes the separation of products and unreacted substrates.

CO_2 gas shows exceptional solubility in a number of ionic liquids, while CO is less soluble and hydrogen is only slightly soluble in such ionic liquids.

Some ionic liquids can be distilled under vacuum at temperature near 300°C [89] despite of their extremely low vapour pressures. It was assumed earlier that the vapour may be comprised of individual, separated ions [90], but it was later on proved that the vapour formed consisted of ion pairs. [91] Thermal stability and melting point of ionic liquids depend on the components of the liquid.

Because of their distinctive properties, ionic liquids are attracting attention of chemists all over the globe, particularly in fields like organic chemistry, electrochemistry, catalysis and engineering (magnetic ionic liquid).

Designer Solvents for a Cleaner World

Ionic liquids have been described as designer solvents [92], as these are made up of at least two components (the anion and cation) which can be varied. The solvents can be designed to posses a particular set of properties by the choice of appropriate cations and anions. Hence, the term "designer solvents" has been justified for ionic liquids.

The melting points of 1-alkyl-3-methyl imidazolium tetrafluoroborates [93] and hexafluorophosphates [94] depends on the length of the 1-alkyl group forming liquid crystalline phase for alkyl chain lengths having more over than twelve (12) carbon atoms. The miscibility of water in these ionic solvents also varies with their structures. 1-alkyl-3-methylimidazolium tetrafluoroborate salts with the alkyl chain length having carbon atoms less than 6 are miscible with water at 25°C but with carbon atoms more than six, a separate phase when it is mixed with water.

As the relative solubilities of the ionic and extraction phase can be adjusted to make the separation relatively easier. This behaviour of ionic liquids can be beneficial for solvent extractions and separations of products. Various ionic solvents are designed different purposes as follows:

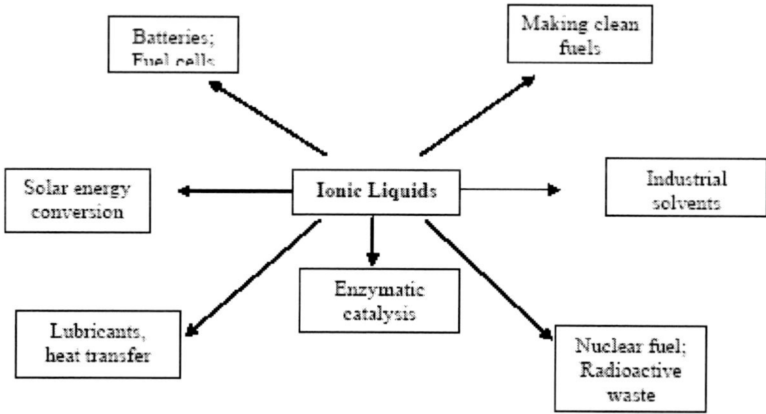

Room Temperature Ionic Liquids

These liquids consist of bulky and asymmetric organic cations such as 1-alkyl-3-methylimidazolium, 1-alkylpyridinium, N-methyl-N-alkylpyrrolidinium and ammonium ions. A wide range of anions may be used from simple halides, which shows high melting points, to complex inorganic anions (tetrafluoroborate and hexafluorophosphate) and large organic anions (bistriflimide, triflate or tosylate).

Some non-halogenated organic anions such as formate, alkylsulphate, alkylphosphate or glycolate are also used. One of the first room temperature ionic liquid was a mixture of [emim]Cl with $AlCl_3$, but it is not water stable. The first water insoluble room temperature ionic liquid is [bmim[[PF_6].

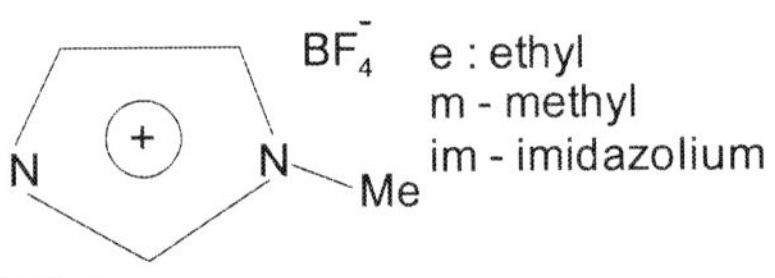

BF_4^- e : ethyl
m - methyl
im - imidazolium

[emim][BF_4]

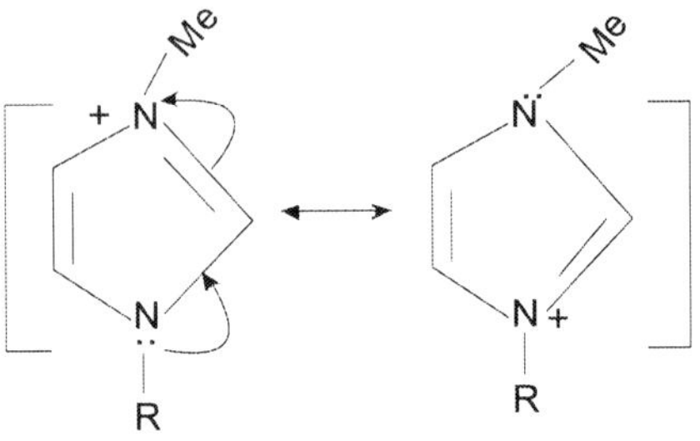

Some other examples are: [emim] [PF_6], [emim] [NO_3] and [emim] [ClO_4]

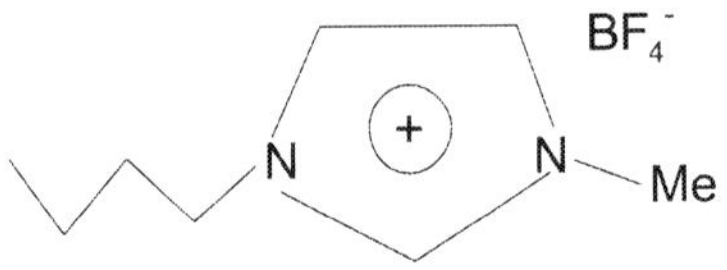

[emim] [BF_4] 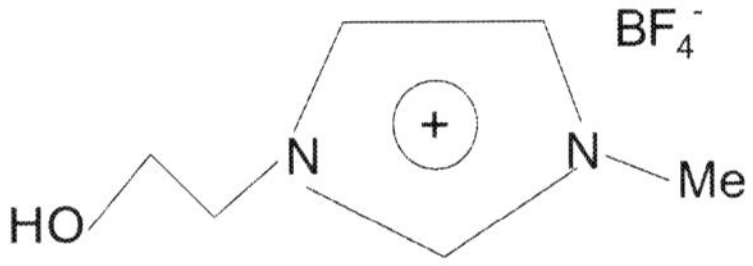 [hyemim] [BF_4]

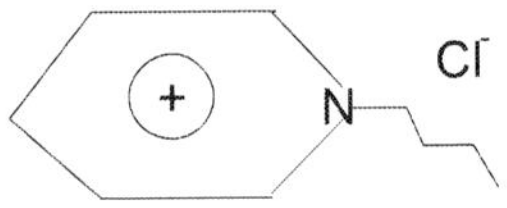

[bpy] [Cl] 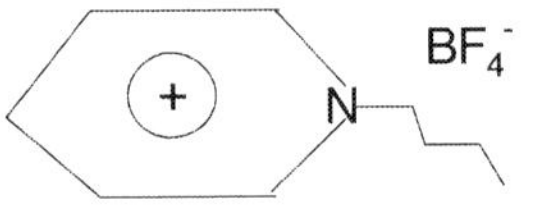 [bmpy] [BF_4]

Reactions in chloroaluminate (III) ionic liquids: $AlCl_3$ is less soluble in many solvents and hence, this ionic liquid promotes reactions that are conventionally promoted by $AlCl_3$. These ionic liquids were prepared by mixing the appropriate organic halide salts with $AlCl_3$, which results in the two solids melting together to form the ionic liquid. A classical reaction promoted by chloroaluminate (III) ionic liquids is the Fridel-Craft's reaction [95] and traseolide (5-acetyl-1,1,2,6-tetramethyl-3-isopropylindane) has been obtained with high yields. [96]

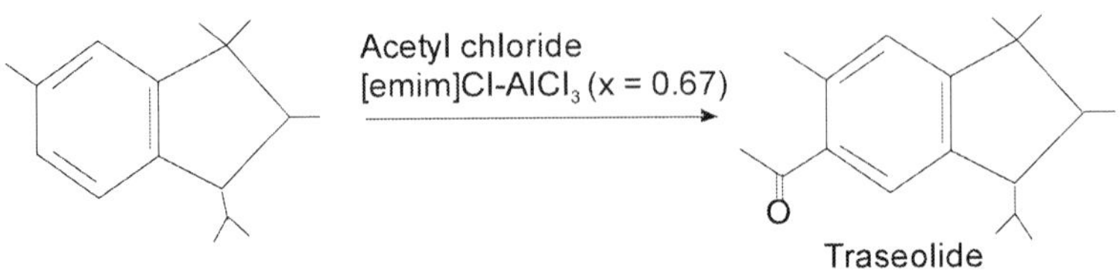

Acetyl chloride
[emim]Cl-$AlCl_3$ (x = 0.67)

Traseolide

The use of ionic liquid gives the highest known selectivity for the 1-position (89%) as compared to 2 position (2%) in the acylation of naphthalene.

Acetyl chloride
[emim]Cl-$AlCl_3$ (x = 0.67)
5 min; 0°C
1-isomer (89%
1-isomer (2%)

Isomerization and cracking reactions of alkanes and alkyl chains in chloroaluminate (III) ionic liquids occurs at temperatures as low as 90°C, whereas conventional catalytic reactions require much higher temperature for these processes; typically in the range of 300-1000°C [97].

Pyrene and anthracene can be reduced to the thermodynamically most stable isomer (perhydroanthracene; 90% yield) at ambient temperature and pressure while catalytic hydrogenation reactions requires high temperature and pressure as well as and an expensive platinum oxide catalyst giving rise to isomeric mixture of products. [98]

The composition of a tetrachloroaluminate (III) ionic liquid is normally given in the form of apparent mole fraction of $AlCl_3$ {x ($AlCl_3$)} present. The ionic liquids with x ($AlCl_3$) < 0.5 contains an excess of Cl^- ions over $[Al_2Cl_7]^-$ ions and these are called "basic" ionic liquids, while those with x ($AlCl_3$) > 0.5 contains an excess of $[Al_2Cl_7]^-$ ions over Cl^- and these are called acidic ionic liquids. The ionic liquids with x($AlCl_3$) = 0.5 are called neutral ionic liquids.

Neutral Ionic Liquids

In spite of chloroaluminate (III) ionic liquids also suffer from several disadvantages; such as moisture sensitivity and difficulty in product separation. In order to develop more chemistry in ionic liquids, it is desirable that water stable ionic liquids are there and hence, these become important.

An example of this kind is [bmim] [PF_6] [99], which forms triphasic mixtures with water and alkanes. This type of a multiphasic behaviour has important implications for clean (green) synthesis. Transition metal catalysis can be exclusively dissolved in the ionic liquids and thus, products and byproducts can easily be separated from the ionic liquid by solvent extraction either with water or an organic solvent. When precious metal catalyst are used with expensive ligands; both ionic liquid as well as catalyst may be recycled and reused again and again.

Volatile products can be easily separated by distillation, as ionic liquids has no vapour pressure and therefore, these can not be lost daily distillation. Secondly, there is no requirement of having any special conditions, when reactions are carried out in ionic solvent like there is no need to exclude water, or to provide an inert atmosphere.

Neural ionic liquids are the best choice as solvents for hydrogenation reactions, because they provides a homogeneous transition – metal catalyst and products can be easily separated. An example of asymmetric hydrogenation is synthesis of (S) – naproxen in ionic liquid. [100]

H_3CO COOH
[$RuCl_2$ - (S) BINAP] / H_2
[bmim] [BF_4]/Pr^iOH
H_3CO H CH_3 COOH

These ionic liquids are excellent solvents for the Diels-Alder reaction [101,102] also as, there is an enhancement as compared to reaction carried out in molecular solvents, including water. [103]

In conventional Heck reactions (palladium – catalysed coupling of aryl halides with alkenes), the catalyst is usually lost at end of the reaction but in ionic liquids, the expensive catalyst can be easily recycled, which is an added advantage of using ionic liquids as a green solvent.

In this reaction, aromatic anhydrides are used as a source of the aryl group and by-product of the reaction is benzoic acid, which can be readily converted back to the anhydride and as a result, the halide containing waste is not formed.

Methylene insertion reactions are carried out easily in ionic liquids in alkaline medium using a chiral sulphide, an asymmetric methylene insertion reactions can be carried out in good enantiomeric excess using [bmim] [PF_6].

O H + [S]$^+$ [I] [bmim] [PF_6] / KOH → O + KI

O H + Br [bmim] [PF_6] / KOH → O
S

Looking to various advantages, the room-temperature ionic liquids have almost replaced the conventional organic solvents in bio-transformations. [104]

CH_3 N H Base Ionic liquid → CH_3 N N O Cl O R Ionic liquid →
Cl N O HCl
O N N O OCH_3

Pravadoline

The complete synthesis of pravadolione in [bmim] [PF_6] has been given by Earle *et al.* [105]

Low temperature ionic liquids: These ionic liquids (below 130 K) have been proposed as the fluid base for spinning liquid mirror telescope to be based on the Earth's moon. [106] Low temperature is advantageous in imaging infrared light and low volatility is important for use in the vacuum conditions present on the moon.

APPLICATIONS

(i) Food science: Ionic liquids have been used in food science. [bmim] Cl is able to completely dissolve freeze dried banana pulp so that the entire banana composition can be monitored as a function of banana ripening. The solution of [bmim] Cl with 15% DM_{50} is also used in carbon-13 NMR analysis.

(ii) Nuclear industry: Room temperature ionic liquids are extensively used for innovative applications in nuclear chemistry. Ionic liquid is also used as extractant in solvent extraction systems, as alternate electrolyte media for the high temperature pyrochemical processing. Recently, a novel extraction processing using ionic liquids was developed and demonstrated for the recovery of palladium. [107]

(iii) Solar energy applications: Ionic liquids have a great potential for heat transfer and storage medium in solar thermal energy systems.

(iv) Hydrogen storage: Due to desired properties of ionic liquids, [bmim]Cl is used in the dehydrogenation of ammonia borane. Immediately upon heating the sample, hydrogen evolution took place as high as 5.4 wt % H_2. [108]

(v) Natural production extraction: Ionic liquids are superior to conventional solvents in the extraction of specific natural compounds from plant biomass for pharmaceutical and cosmetic applications. For example, protic ionic liquids have been used as solvents for the isolation of the antimalarial drug artimisinin from the plant *Artomisia annua.*

Due to their non-volatility, which effectively eliminates a major pathway for environmental pollution and contamination, ionic liquids have been considered as having a very low impact on the environment and human health, and are thus, recognized as solvents for green chemistry.

GREEN REAGENT

In any synthesis of a target molecule, the selection of an appropriate reagent is quite necessary. This selection is made on the basis of efficiency, availability and its effect on the environment. The selection of a particular reagent versus another reagent for the same transformation can affect the nature of by-products, percentage, yield, time taken etc. Following are some of the examples of green reagents.

a) Dimethylcarbonate (DMC): DMC is a non-toxic and environmentally safe reagent, which can be used in organic synthesis as a green substitute for toxic intermediates. Tundo [109,110] developed a method to methylate active methylene compound using DMC, where no inorganic salts are produced.

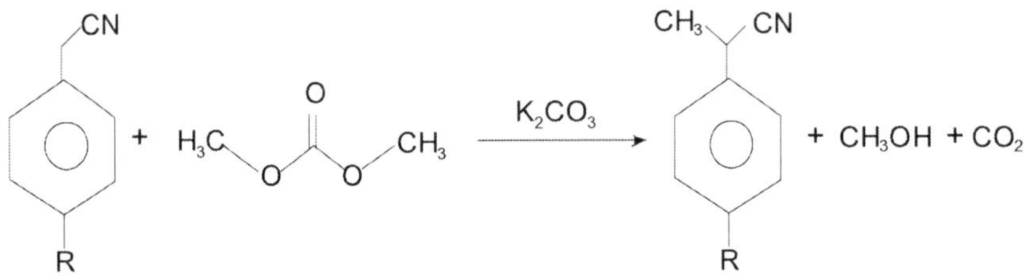

DMC is also used as an organic oxidant for e.g. cyclopentanone react with DMC and a base K_2CO_3 to yield adipic acid.

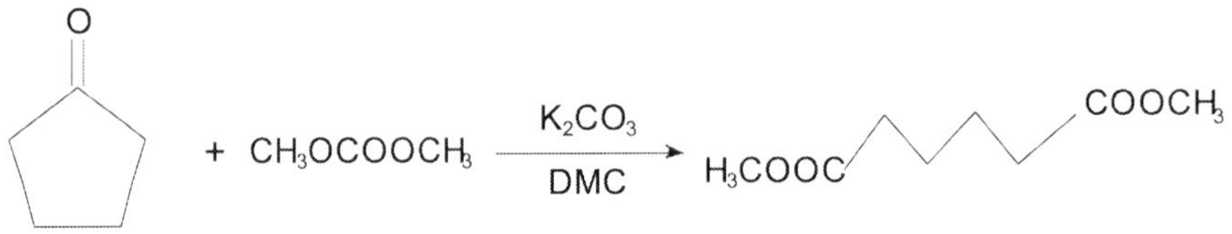

b) Polymer supported reagents: Polymer supported reagents are those, in which ordinary reagents are bound to some polymer support. Supported reagents are usually used for clean and smart synthetic procedure because it allows reaction to be performed more selectively, more efficiently and it can be repeated. Some of such reagents are as follows:

(i) Polymer supported peracids: These reagents are used for the epoxidations of alkenes with good yield. [111]

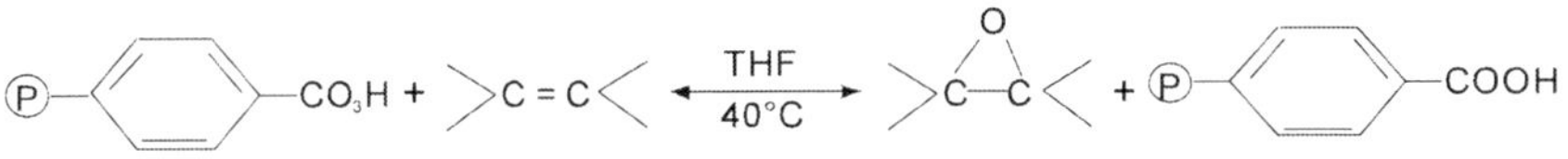

(ii) Polymer supported chimeric acid: Chromic acid is a known oxidizing agent. The major use of chromic acid in synthetic chemistry is the oxidation of primary and secondary alcohols to corresponding aldehydes and ketones, respectively. [112,113] Commercial available form of this reagent is Amberlyst A-26, $HCrO_4^-$. The advantages for using polymer supported oxidizing agent is that it decreases side reactions and the oxidation process stops at the product aldehyde and ketone only and not does proceed to yield acids.

(iii) Poly-N-bromosuccinimide (PNBS): The reagent is an efficient brominating agent and is used as benzylic and allylic brominating agent. The structure of PNBS is –

Cumene on bromination [114] gives α,β,β' –tribromocumeme.

PNBS / CCl_4

Cumene → α,α,β'-Tribromocumene

(iv) Polystyrene Wittig reagent: The polystyrene Witting reagent [115] can be synthesized by this process.

$ClPPh_2$, $RR'CHBr$, $N\text{-}C_4H_9Li$

Polytyrene Wittig reagent

This reagent reacts with carbonyl compounds to yield corresponding alkenes.

Alkene

(v) Polymeric thioonisolyl resin: Polystyrene methyl sulphide on reaction with chlorine gives S-chloro sulphonium chloride resin in the presence of trimethylamine and it is used for selective oxidation of the the alcohols. [116]

Cl_2 / Et_3N

S-chloro sulphonium chloride

$HO\text{-}(CH_2)_n\text{-}OH$

$HO(CH_2)_{n\text{-}1}CHO$ (50.2%)

+

$OHC(CH_2)_{n\text{-}2}CHO$ (2.2%)

(vi) Polymeric phenylthiomethyl lithium reagent : This reagent is useful for chain lengthening of alkyl iodide with good yield.

$$\text{Ⓟ}-C_6H_4-SCH_2^- Li^+ + ICH_2(CH_2)_nR \longrightarrow \text{Ⓟ}-C_6H_4-SCH_2(CH_2)_{n+1}R$$

$$\xrightarrow{NaI / CH_3I \;|\; DMF} R(CH_2)_{n+1}CH_2I + \text{Ⓟ}-C_6H_4-SCH_3$$

(vii) Polystyrene anhydride: Polystyrene anhydride is efficiently used for acetylation of aniline with 90% yield. [117] Similarly, acylation of ethanol gives ethyl benzoate in 90% yield.

(viii) Polystyrene carbodiimide: This reagent is useful in Moffatt oxidation of alcohols to their corresponding aldehydes and ketones. [118] It is also used for the synthesis of anhydrides. [119]

(ix) Polymer supported peptide coupling agent: Polystyren-1-ethyl-3-(3'-dimethylamino propyl)- carbodiimide hydrochloride (Ps –EDCI) used as a peptide coupling reagent. [120] It mediates the formation of amides through the coupling of amines with carboxylic acids. Polymer supported ethyl-1,2-dihydro-2-ethoxy-1-quinolinecarboxylate (EEDQ) i.e., PS – EEDQ is also used for forming peptide linkage. [121]

Green Catalyst

Catalyst plays a major role in establishing the economic strength of any chemical industry, because the use of a catalyst facilitates transformation without being consumed in the reaction and without being incorporated in the final product. It can be recycled also in most of the cases.

Types of catalyst:

a) Acid catalysts: Hydrogen fluoride is an extremely corrosive, hazardous and toxic catalyst which was used in the production of linear alkylbenzenes. It has now been replaced by a solid acid catalyst, which does not require special material of construction and it also has lower operating costs.

Microencapsulated Lewis acids have replaced traditional corrosive monomeric Lewis acids in Michael [122], Friedel-Crafts [123], Mannich [124], Iminoaldol [125] reactions.

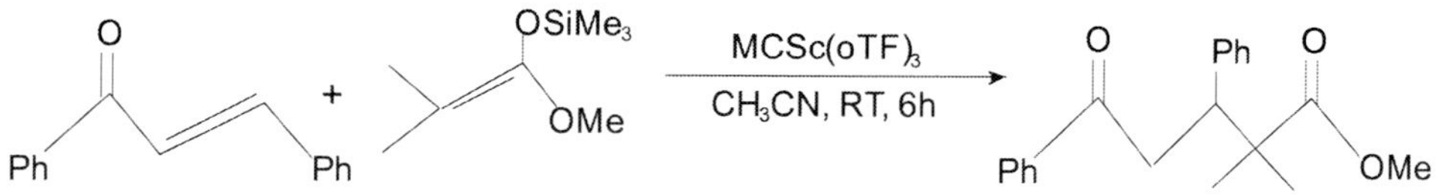

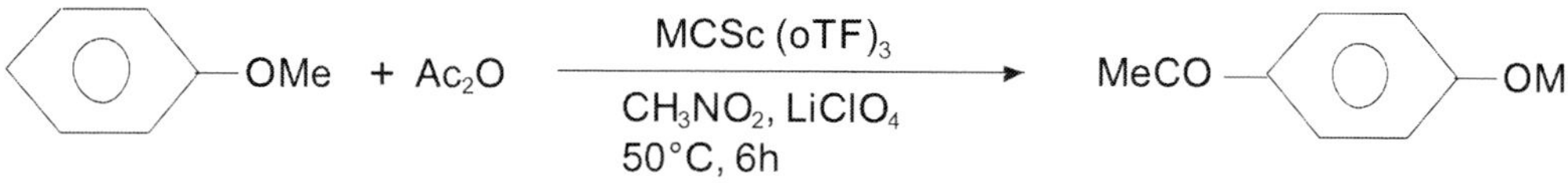

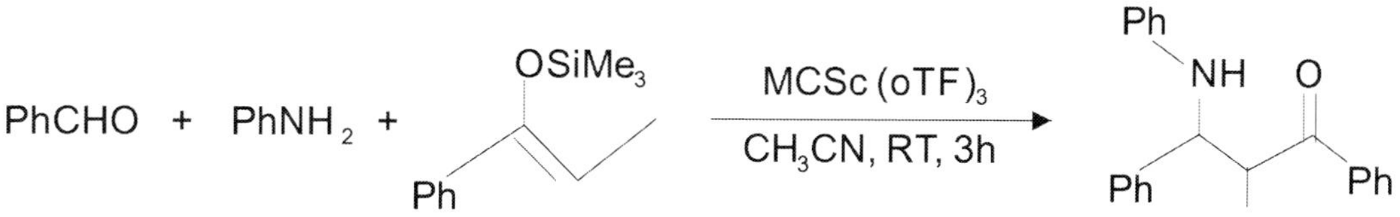

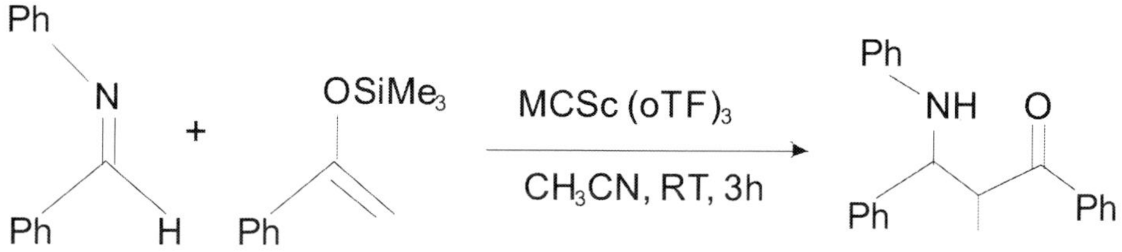

[MCSc (oTF)$_3$ = Microencapsulated scandium trifluoro methane sulphonate

Zeolties are quite widely used in petrochemical industries in acid catalysed processes. [126,127]

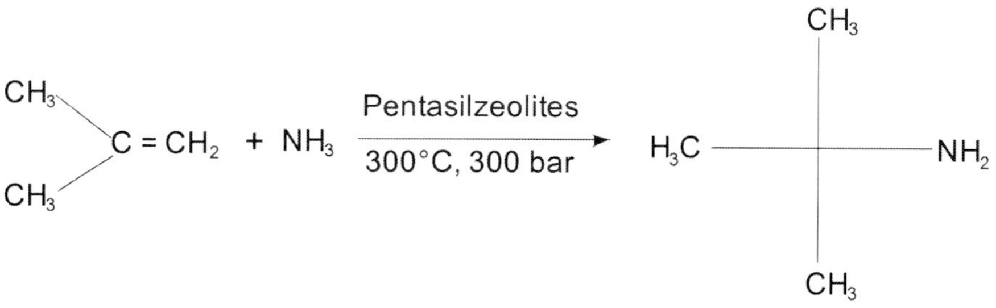

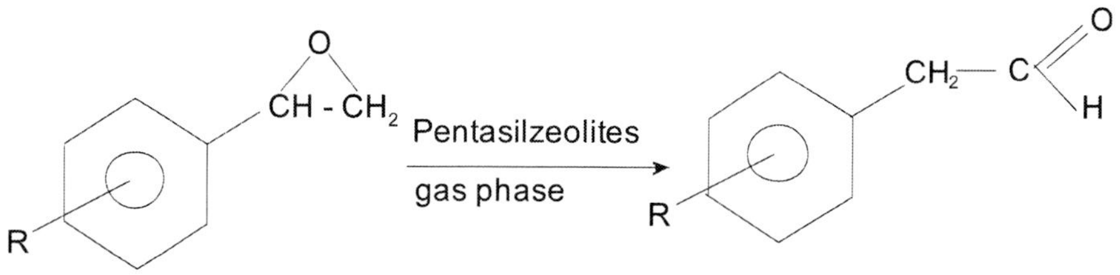

Clayzic is also used as a acid catalyst. The pores of clayzic favour the desire intramolecular cyclisation over polymerization [128] and olefination of benzaldehyde. [129]

R

$CH_2CH(OA)_2$

S

Clayzic

PhCl, 132°C

R

O

+

EPZ10

$MeNO_2$, RT

b) Oxidation catalysts: Titanium and vanadium silicates are also used as a oxidation catalysts. [130] Titanium silicates (TS) are used in hydroxylation of phenol. This is a clean process, which gives an excellent conversion phenol to hydroquione with very little waste.

OH

TS-1

H_2O_2

OH + OH + OH OH

OH OH

87% 7% 12%

Vanadium silicates are also capable of selectivity oxidizing 4-chlorotoluene. [131]

CH_3

Cl

VS-2/H_2O_2

373 K/MCCN

CHO + CH_2OH + CH_3

OH OH

Cl Cl Cl

66% 21% 8%

Oxidation catalyst is based on chemically modified support materials include Co, Cu and Fe also. [132]

c) Basic catalysts: Basic catalysts are used in the alkylation of phenol, side chain alkylation and isomerization reactions. [133,134]

$$\text{Phenol} + CH_3OH \xrightarrow{MgO} \text{2,6-dimethylphenol}$$

$$+ \xrightarrow[\gamma - Al_2O_3]{KOH/K}$$

$$\xrightarrow[\gamma - Al_2O_3]{NaOH/Na} \quad +$$

d) Polymer supported catalysts: Conventional catalysts are linked to a polymer backbone and are used in this form to catalyse different reactions.

(i) Polystyrene-aluminium chloride: Reactions, where both; a dehydrating agent and a Lewis acid are required, then polystyrene-$AlCl_3$ is a useful catalyst there. It is used to prepare ethers from alcohols. [135]

$$H-C-OH \xrightarrow{P-C_6H_4-AlCl_3} H-C-O-C-H$$

Dicyclopropylcarbinol — Di (dicyclopropylcarbinyl) ether

(ii) Polymeric super acid: It is AlC13 bound with sulphonated polystyrene. [136,137] Normally, the cracking and isomerization of alkanes is carried out in the presence of Lewis acid at high temperatures and high pressures but in the presence of a polymeric super acid catalyst. Such reactions occur at 375°C and atmospheric pressure.

(iii) Polystyrene metalloporphyrins : These catalysts [138] facilitate the oxidation of thiols to disulphide even at room temperature.

$$RSH + \text{base} \rightleftharpoons RS^- + H^+ + \text{base}$$

$$2\,RS^- + O_2 \longrightarrow 2\,RS^- + O_2^-$$

$$2\,RS^{\bullet} \longrightarrow R_2S_2$$

$$O_2^- + H_2O \longrightarrow 2\ OH^- + \tfrac{1}{2}\ O$$

(iv) Polymer supported photosensitizers: The photosensitizers supply molecular oxygen in a photochemical reaction. The efficiency of polymer-bound rose-bengal photisensitizer (SENSITOX) is about 65% of the rose-bengal but the high yield of product and the ease of isolation compensate the slightly longer reaction period.

Rose Bengal

DMF 60°
20 hr

(v) Polymer supported phase transfer catalysts: The onium salts immobilized on a polymer matrix are slower in action, because it is difficult to bring the reactants in contact with the catalytic site. This drawback is overcome by using a long chain to bind the catalyst to the polymer matrix. By using these catalysts one can separate product more easily and recycle the catalyst also. 1-cyanooctante can be prepared by stirring a mixture of 1-bromooctane and potassium cyanide in the presence of resin P-$(CH_2)_6$-P^+ $(C_4H_9\text{-n})_3$ Br^- for 1.6 hrs. [139]

$$n\text{-}C_8H_{17}Br \xrightarrow[\text{(P)-PTC}]{KCN} n\text{-}C_8H_{17}CN$$

1-Bromooctane 1-Cyanooctane (80%)

(vi) TiO_2 photocatalyst in green chemistry: TiO_2 based photocatalytic systems are important for the purification of polluted water [140-145], decomposition of toxins as well as offensive atmospheric odors and chlorofluorocarbons on a huge global scale.

(vii) Solid supported reagents: [146] Some examples of these are:

- Cyanoethylation of alcohols and thioalcohols:

R - OH + $CH_2=CH-CN$ —Hydrotolcite, CH_2Cl_2, RT→ R–O–CH_2CH_2–CN

- Selective synthesis of β-nitroalkanols:

RCHO + CH_3NO_2 —Mg-Al-hydrotalice→ R–CH(OH)–CH_2NO_2

- Acylation of alcohols with carboxylic acids:

Clay catalyst

R – OH + AcOH ⟶ R – oAC + ROR + H_2O

- Formation of C-C bond:

R(R')C=O + CH_2(Y)CN —LDH-F→ R(R')C=C(CN)Y + H_2O

LDH-F is a layered double hydroxide fluoride – solid base catalyst.

(viii) Synthesis of bromoorganics: These are important precursors for the preparation of many pharmaceuticals and agrochemicals. [147] The preparation of these compounds involve Br_2, which is toxic in nature. Some newer and ecofriendly bromination and brominating agents have been reproted. [148]

V_2O_5

3 Bu_4 $NHBr_3 + H_2O_2$ ⟶ Bu_4NBr_3 + 2 Br_4NOH

ArH + Bu_4NBr_3 ⟶ Ar=Br + Bu_4NBr + HBr

$R_2C=CR_2$ + Bu_4NBr_3 ⟶ $R_2CBr–CBrR_2$ + Bu_4NBr

(ix) Synthesis of isooctane: Isooctane is an integral part of gasoline. Solid acids like alumina, zeolites, etc. act as catalyst in the synthesis of isooctane. [149]

$$CH_3-\underset{CH_3}{\overset{CH_3}{|}}-H \; + \; \text{(butene)} \xrightarrow[\text{Solid acid}]{Br} CH_3-\underset{CH_3}{\overset{CH_3}{|}}-CH_2-\underset{CH_3}{CH}-CH_3$$

Isooctane

Supercritical Fluids (SCF): Supercritical fluids (SCFs) are intermediate between liquids and gases.

Supercritical carbon dioxide ($ScCO_2$): CO_2 shows a large change in its solvent properties for a relatively small change in pressure. It has high compressibility and high range of solvent properties. $ScCO_2$ with small amount of co-solvents can further modify its solvent properties. $ScCO_2$ has a potential for increasing reaction rates, because of its high diffusion rates and it has also potential for homogeneous catalytic processes. High solubility of light gases, some catalysts and substrates in $ScCO_2$ bring them all together in single homogeneous phase. CO_2 is the last product of oxidation and hence, it provides an excellent medium for oxidation and reduction reactions. Its high specific heat capacity ensures efficient heat transfer in mostly highly exothermic oxidation reactions.

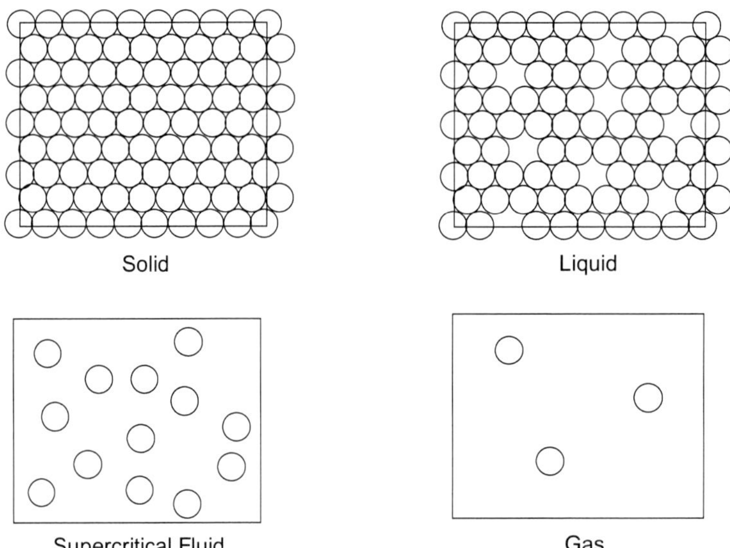

The oxidation of sulphides to sulphoxides is a quite commonly used transformation in organic chemistry. Oakes *et al.* [150] have developed conditions to carry out this reaction in $ScCO_2$, with clean sulphoxide formation with their overoxidation to the sulphone. By modification of the density, the selectivity can be enhanced to a maximum diastereoselectivity of $> 95\%$ While no diastereoselectivity is observed in conventional solvents. Sparks *et al.* [151] studied supercritical-fluid-assisted oxidation of oleic acid with ozone and potassium permanganate.

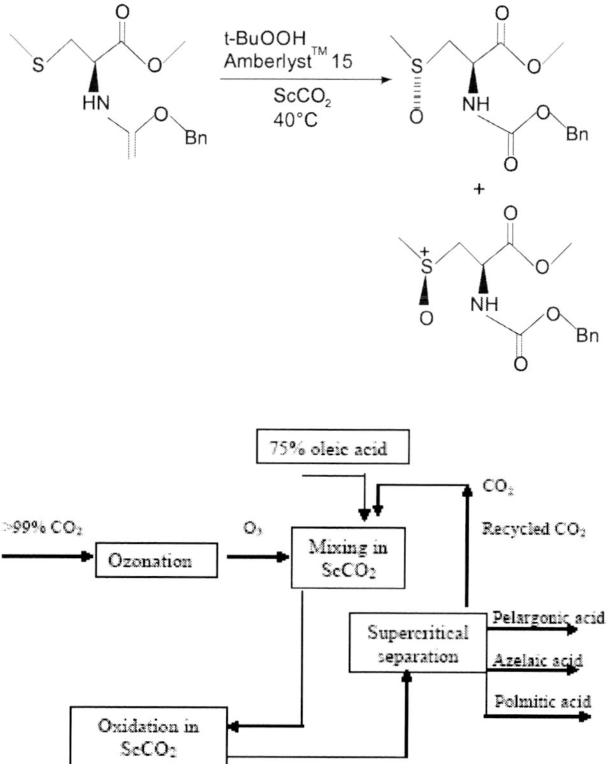

Flow diagram for ozonylsis of oleic acid

(2-Percarboxyethyl)-functionalized silica performs the Baeyer-Viliger oxidation of ketones in $ScCO_2$ at 250 bars and 40°C. The reagent can be recycled by treatment with acidic 70% hydrogen peroxide at 0°C. [152]

Karmee *et al.* [153] reported Pd (O) catalysis enantioselective sullphoxidation by cascade reaction. $ScCO_2$ was used as a medium for *in situ* generation of H_2O_2 directly from H_2 and O_2.

ScCO2
H2
O2
Pd(O)
H2O
H2O2
CPO

Wang *et al.* [154] studied polyethylene glycol radical initiated oxidation of benzylic alcohols in compressed CO_2. It is a cost effective and environmentally benign pathway, in which PEG, O_2 and CO_2 act as initiator, oxidant and solvent, respectively without the need of a catalyst and additive.

OH

PEG / O_2/ CO_2

120°C, 16MPa, 12h

O

90%

CH_2OH

PEG / O_2/ CO_2

120°C, 16MPa, 12h

O

Oh

89%

Reactions usually carried out in chlorofluorocarbon solvents can also be carried out in liquid or $ScCO_2$.

Thus, $ScCO_2$ may be used as a reaction medium for hydrogenation, hydroformylation, epoxidation, radical reactions, Pd-mediated C-C bond formation, ring closing metathesis, biotransformations and in polymerization. [155]

Problems in using $ScCO_2$:

- Moderate pressure is required, but in laboratory standard HPLC apparatus is used.
- It is a weak solvent and relatively non-polar so that the use of co-solvents is required.
- Reaction in the presence of good neucleophiles is often reversible.
- Energy is required for compression of CO_2 and thus, energy consumption reduced recycling and decompression.

The generation of benign alternate synthetic routes is the need of the day and tomorrow. One should not assume that the presently available synthetic routes are the only one possible route and therefore, there is a pressing demand to find newer routes, which create little or no hazardous wastes.Thus, we should accept these challenges placed before the chemistry community, which are posed by already polluted and further polluting environment in coming millennium and come out with some benign chemistry so as to add a new feather in the cap of the chemists in general. This success will also act as a catalyst in stimulating many more chemists to select this field of study and develop some eco-friendly chemical routes so that the extent of pollution on the earth decreases and the humanity feels relieved of the strong grip of the environmental pollution. So that with pride, we can say –

Green Chemistry : Green Earth,
Clean Chemistry : Clean Earth [156]

REFERENCES

[1] Paul T. Anastas and John C. Warner, *Green Chemistry : Theory and Practice,* Oxford University Press, New York, 1998.

[2] Barry M. *Trost, Science*, 1991, 1471.

[3] K. Takai and C. N. Heathcock, *J. Org. Chem*., 1985, 50, 5247.

[4] A. E. Vouioukas and H. B. Kagon, *Tetrahedron Lett.,* 1987, 28, 5513.

[5] A. Lubineau, *J. Org. Chem*., 1986, 56, 2142.

[6] A. Lubineau and E. Mayer, *Tetrahedron*, 1988, 44, 6065.

[7] S. Kobayashi and I. Hachiya, *J. Org. Chem*., 1994, 59, 3590.

[8] R. Breslow, *Acc. Chem. Res*., 1991, 24, 159.

[9] P. A. Grieco (review), *Aldrichim. Acta.,* 1991, 24, 59.

[10] D. C. Rideout and R. Bresow, *J. Am. Chem. Soc*., 1980, 102, 7816.

[11] B. Woodward and H. Baer, *J. Am Chem. Soc.,* 1948, 70, 1161.

[12] W. Oppolxer, Angew. *Chem. Int. Ed. Engl.,* 1972, 11, 1031.

[13] P. –S. Kwan, Y. –M. Kim, C. –J Kang, T. –W. Kwan, Synth. *Commun.,* 1997, 27, 4091.

[14] K. Takai and C. H. Heathcock, *J. Org. Chem.,* 1985, 50, 3247.

[15] K. Mikami, M. Terada and T. Nakai, *J. Chem. Soc. Chem. Commun*., 1993, 343.

[16] M. Costa, F. Areias, L. Abrunhosa, A. Venancio and F. Proenca, *J. Org. Chem*., 2008, 73, 1954.

[17] A. J. Lapworth, *J. Chem. Soc*., 1903, 83, 995.

[18] D. Enders, O. Niemeier and T. Balensiefer, *Angewandte Chemie International* Edition, 2006, 45, 1463.

[19] E. T. Kool and R. Breslow, *J. Am. Chem. Soc.,* 1988, 110, 1596.

[20] K.-I. Iwamoto, M. Hamaya, N. Hashimoto, H. Kimura, Y. Suzuki and M. Sato, *Tetrahedron Lett.,* 2006, 47, 7175.

[21] A. I. Vogel, *Textbook of Practical Organic Chemistry Including Qualitative Analysis, Longmans*, London, 1988.

[22] R. H. Dodd and M. L. Hyaric, *Synthesis,* 1993, 295.

[23] Y. J. Kim and R. S. Verma, *J. Org. Chem*., 2005, 70, 7882.

[24] M. L. Pound, A. Loupy, S. Maraue and A. Pett., *Heterocycles,* 2004, 63, 297.

[25] S. Surveswari and T. K. Raja, *Indian J. Chem*., 2006, 45B, 546.

[26] C. M. Kormos and N. E. Leadbeater, *J. Org. Chem*., 2008, **73**, 3854.

[27] I. Rukhsana, I. A. Kureshy, K. Pathak, N. H. khan, Sayed, H. B. Abdi, H. C. Bajaj and E. Suresh, *Research Lett. Org. Chem*., 2009, 5.

[28] Y. Wef and R. Bakthavatechalan, *Tetrahedron Lett.,* 1991, 32, 1535.

[29] P. Salehi, N. M. Khodaei, M. A. Zolfigol, A. Keyvan, *Monatshefte Fur Chemie., 2*002, 133, 1291.

[30] D. J. Upadhyay and S. D. Samant, *Appl. Catal. A: Gen.,* 2008, 340, 42.

[31] F. Toda, K. Tanaka, Y. Kagowa and Y. Sakaino, *Chem. Lett.,* 1990, 373.

[32] I. Almera, A. Diaz-Ortiz, E. Diez-Barra, A. *Hos and A. Loupy, Chem. Lett*., 1996, 333.

[33] W. S. Bremner and M. G. *Organ, J. Comb. Chem*., 2007, 9, 14.

[34] R. D. Virre, G. Evindar and R. A. Batey, *J. Org. Chem*., 2008, 73, 3452.

[35] W. J. Hao, B. Jiang, S. J. Tu, X. D. Cao, S. S. Wu, S. Yan, X. H. Zhang, Z. G. Han and F. Shi, *Org. Biomol. Chem*., 2009, 7, 1410.

[36] Shui-Teinchem, Shyh-Horng Chiou and Kung-Tsungwang, *J. Chem. Soc. Chem. Commun.*, 1990; 807.

[37] M. N. Gedye, F. E. Smith and K. C. Westaway, *Can. J. Chem.*, 1988, 66, 17.

[38] R. S. Verma and R. W. Saini, *Tetrahedron Lett.*, 1997, 38, 2623.

[39] R. S. Verma and D. Kumar, *Synth. Commun.*, 1999, 29, 1333.

[40] R. N. Gedye, W. Rank and K. C. Westaway, *Can. J. Chem.*, 1991, 69, 700.

[41] P. Guang, Y. Xue, P. Guang, X. Fen, 2009, 29, 428.

[42] J. Gulin and G. Guncheng, *Synth. Commun.*, 1994, 24, 1045.

[43] G. B. Jones and B. J. Chapman, *J. Org. Chem.*, 1994, 58, 5558.

[44] G. Opitz and E. Tempel, *Liebigs Ann. Chem.*, 1966, 699, 68.

[45] R. J. Giguene, T. L. Bray and S. M. Duncan, *Tetrahedron Lett.*, 1986, 27, 4945.

[46] A. Commariev, W. Hoeiderich, J. A. Laffitte, M. P. Dupont, *J. Mol. Catal. A: Gen.*, 2002, 182, 137.

[47] Z. Minoru, *J. Pestic.* Sci., 2006, 31,1.

[48] A. Stobrawe, P. Makaczyk, C. Maillet, J.-L. Muller, W. Leitner, *Angewandte Chemie International Edition*, 2008, 47, 6674.

[49] O. Marvi, D. Habibi, *Chinese J. Chem.*, 2008, 26, 522.

[50] F. Jam, m. Tullberg, K. Luthman, M. Grotti, *Tetrahedron*, 2007, 63, 9881.

[51] T. Kurz, M. Khankischpur, K. Widyan, *Tetrahedron Lett.*, 2006, 47, 4241.

[52] A. Pawetczyk, L. Zaprutko, *European J. Med. Chem.*, 2009, 44, 3032.

[53] L.-J. Li, Y.-X. Song, Y.-S Gao, Y—F Li and J.-F. Zhang, E- *J. Chem.*, 2006, 3, 164.

[54] R. S. Verma, A. K. Chatterjee and M. Varma, *Tetrahedron Lett.*, 1993, 34, 4603.

[55] H. G. Forey and D. R. Dalton, *J. Chem. Soc. Chem. Commun.*, 1983, 628.

[56] D. Villemin, M. Lalaoui and A. B. Alloum, *Chem. Ind.*, 1991, 176.

[57] T. J. Mason and J. P. Lorimer, *Applied Sonochemistry*, 2003.

[58] B. P. Joshi, A. Sharma and A. K. Sinha, *Tetrahedron*, 2005, 61, 3075.

[59] G. H. Mahdavinia, S. Rostamizadeh, A. M. Amani, Z. Emdadi, *Ultrasonics Sonochem.*, 2009, 16, 7.

[60] D. S. Krisol, H. Klotz and R. C. Parker, *Tetrahedron Lett.*, 1981, 22, 407.

[61] J. Elguero, P. Goya, J. Lissavestzky and A. M. Valdemillos, C. R. *Acad. Sci.* Paris, 1984, 298, 877.

[62] S. Piiskop, H. Hagu, J. Jarv, S. Salmer, A. Tuulmets, *Proc. Estonian Acad. Sci. Chem.*, 2007, 56, 199.

[63] K. S. Suslik and D. J. Casadonte, *J. Am. Chem. Soc.*, 1987, 109, 3459.

[64] T. Kitazumi and N. Ishikawa, *J. Am. Chem. Soc.*, 1985, 107, 5186.

[65] P. Boudjouk, D. P. Thompson, W. H. Ohrborm and B. H. Hans, *Organometallics*, 1986, 5, 1257.

[66] M. H. Entezari and A. A. Shameli, *Ultrasonics Sonochem.*, 2000, 7, 169.

[67] J. L. Luche and J. C. Damianu, *J. Am. Chem. Soc.*, 102, 7926.

[68] J. Menedez, G. G. Trigo and M. M. Solhuber, *Tetrahedron Lett.*, 1986, 27, 3285.

[69] S. Yamazaki, T. Mori and T. Katou, *J. Photochem. Photobiol.* 2008, 199A, 330.

[70] M. E. Sigman, J. T. Barbas, S. Corbett, Y. Chen, I. Lavanov, R. Dabestani, *J. Photochem. Photobiol.* 2001, 138 A, 269.

[71] S. C. Ameta, R. Ameta, A. Duggar and A. Kumar, *Maced.* J. *Chem. & Chem. Engg.*, 2009, 28.

[72] M. Kurtz, O. Hiririchseno, R. Becker, S. Rabe, K. Merz and Orie, *Catal. Lett.,* 2004, 92, 49.
[73] M. A. Centeno, M. C. Hidalgo, M. I. Dominguez, *Catal. Lett.,* 2008, 123, 198.
[74] R. Podsiadly, *J. Photochem. Photobiol.*, 2008, 198A, 60.
[75] M. G. Kang, H. S. Jung and K. J. Kin, *J. Photochem. Photobiol.*, 2004, 162A, 423.
[76] O. Sijbren, and J. B. F. N. Engberts, *Org. Biomol. Chem.,* 2003, 1, 2809.
[77] M. Biscoe and R. Breslow, *J. Am. Chem. Soc.,* 2003, 125, 12178.
[78] R. Breslow, *Acc. Chem. Res.,* 2004, 37, 471.
[79] R. Breslow, K. Groves and M. U. Mayer, *J. Am. Chem. Soc.*, 2002, 124, 3622.
[80] M. Biscoe and R. Breslow, *J. Am. Chen. Soc.,* 2005, 127, 10812.
[81] Y. Cheng, H. Fan, S. Wu, Q. Wang, J. Cuo, L. Gao, B. Zong and B. Han, *Green Chem.,* 2009, 11, 1061.
[82] B. Baruwati, V. Polshettiwar and R. S. Verma, *Green Chem.*, 2009, 11, 926.
[83] T. Welton, *Chem. Rev.,* 99, 2071 (1999).
[84] J. Holbrey and K. R. Seddo, *Clean Prod. Proc.*, 1, 223 (1999).
[85] K. R. Seddon, *J. Chem. Tech. Biotech.*, 68, 351 (1997).
[86] P. Walden, Bull. *Acad. Imper. Sci.* (St. Petersbreg) 1800 (1914).
[87] H. L. Chum, V. R. Koch, L. L. Miller and R. A. Osteryoung, *J. Am. Chem. Soc.*, 97, 3264 (1975).
[88] J. S. Wilkes, J. A. Levisky, R. A. Wilson and C. L. Hussey, *Inorg. Chem.,* 21, 1236 (1982).
[89] M. J. Earle, J. Esperanca, M. A. Gilea, J. N. Ca Lopes, L. Rebelo, J. W. Magee, K. R. Seddon and J. A. Widegren, *Nature,* 439, 831 (2006).
[90] P. Wasserschied, *Nature,* 439, 797 (2006).
[91] J. P. Armstrong, C. Hurst, R. G. Jones, P. Licence, K. R. J. Lovelock, C. J. Satterly and I. J. Villar-Garcia, *Phy. Chem. Chemi. Phys.,* 9, 982 (2007).
[92] M. Freemantle, *Chem. Eng. News*, 76, 32 (1998).
[93] J. D. Holbrey and K. R. Seddon, *J. Chem. Soc., Dalton Trans*., 2133 (1999).
[94] C. M. Gordon, J. D. Holbrey, A. R. Kennedy and K. R. Seddon, *J. Mater. Chem.,* 8, 2627 (1998).
[95] J. A. Boon, J. A. Levisky, J. L. Pflug and J. S. Wilkes, *J. Org. Chem.,* 51, 480 (1986).
[96] C. J. Adams, M. J. Earle, G. Roberts and K. R. Seddon, *Chem. Commun.*, 2097 (1998).
[97] R. W. J. Westerhout, J. Waanders, J. A. M. Kuipers and W. P. Van Swaaij, *Ind. Eng. Chem. Res.,* 1998, 37, 2293.
[98] D. K. Dalling and D. M. Grant, *J. Am. Chem. Soc.,* 1974, 96, 1827.
[99] J. D. Huddleston, H. D. Willauer, R. P. Swatloski, A. E. Visser and R. D. Rogers, *Chem. Commun*., 1998, 1765.
[100] A. L. Monteiro, F. K. Zinn, R. F. deSouza and J. Dupont, *Tetrahedron Assymetry,* 1997, 177.
[101] M. J. Earle, P. B. McCormac and K. R. Seddon, *Green Chem*., 1999, 1, 23.
[102] T. Fisher, A. Sethi, T. Welton and J. Woolf, *Tetrahedron Lett*., 1980, 40, 793.
[103] D. C. Rideout and R. J. Breslow, *J. Am. Chem. Soc.,* 1980, 102, 7816.
[104] S. G. Cull, J. D. Holbrey, V. Vargas-Mora, K. R. Seddon and G. T. Lye, *Biotechnol. Bioeng*., 2000, 9, 227.
[105] M. J. Earle, P. B. McCormac and K. R. Seddon, *Green Chemistry* (in Press)

[106] E. F. Dorra, O. Seddiki, A. Angel, D. Eisentein, P. Hickson, K. R. Seddon and S. P. Warden, *Nature*, 2007, 447, 979.
[107] M. Jayakumar, K. A. Venkatesan T. G. Shrinivasan and P. R. Vasudeva Rao, *J. Applied Electrochem.*
[108] Karkamkar, C. Aardahl and T. Autrey, *Mat. Matters*, 2007, 2, 6.
[109] P. Tundo, F. Trotta and G. Moraglio, *J. Org. Chem.*, 1987, 52, 1300.
[110] P. Tundo, F. Trotta and G. Moraglio, *J. Chem. Soc., Perkin Trans.*, 1, 1989, 1070.
[111] C. R. Harrison and P. Hodge, *J. Chem. Soc. Chem. Commun.*, 1974, 1009.
[112] Anita L., Jawanjal and Nandini, P. Hilage, *Trans. Metal Chem.*, 2005, 30, 290.
[113] G. Cainella, G. Cardillo, M. Orena and S. Sandri, *J. Am. Chem. Soc.*, 1976, 98, 6737.
[114] C. Yaroslavsky, A. Patchorink and E. Katehalski, *Tetrahedron Lett.*, 1970, 3629.
[115] M. J. Farrall and J. M. J. Frechet, *J. Org. Chem.*, 1976, 41, 3877.
[116] G. A. Crossby, N. M. Weinshenkar and H. S. Lin, *J. Am. Chem. Soc.*, 1975, 97, 2232.
[117] M. B. Stambhu and G. A. Digens, *Tetrahedron Lett.*, 1983, 1627.
[118] K. E. Pfitzner, J. G. Moffatt, *J. Am. Chem. Soc.*, 1963, 85, 3027.
[119] N. M. Weinshenkar and C. M. Shen, *Tetrahedron Lett.*, 1972, 3281.
[120] M. C. Desai and L. M. Stramiellllo, *Tetrahedron Lett.*, 1995, 34, 7685.
[121] J. Brown and R. E. Williams, *Can. J. Chem.*, 1971, 49, 3764.
[122] S. Kobayashi, I. Hachiya, H. Ishitani and M. Araki, *Synlett.*, 1993, 472.
[123] A. Kawada, S. Mitamura and S. Kobayashi, *Synlett.*, 1994, 545.
[124] S. Kobayashi, M. Araki and M. Yasuda, *Tetrahedron Lett.*, 1995, 36, 5773.
[125] S. Kobayashi, M. Araki, H. Ishitani, S. Nasgayama and S. Hachiya, *Synlett.*, 1995, 233.
[126] Corma, *Chem. Rev.*, 95, 559 (1995).
[127] Corma and H. Garcia, *Catal. Today*, 1997, 38, 257.
[128] P. D. Clark, A. Kirk and J. G. K. Yee, *J. Org. Chem.*, 1995, 60, 1936.
[129] H. P. Van Shaik, R. J. Vjin and F. Bickelhaupt, Angew. *Chem. Int. Ed. Engl.*, 1994, 33, 1611.
[130] W. Ramaswamy, S. Sivashanker and P. Ratnasamy, *Microporous Mater.*, 1994, 2, 451.
[131] T. Selvam and A. P. Singh, *J. Chem. Soc. Chem. Commun.*, 1995, 883.
[132] J. Butterworth, J. A. Clark, S. J. Barlow and T. W. Bastock, *U.K. Patent Appl.*, 1996.
[133] J. H. Clark, A. P. Kybett and D. J. Macquarrie, *Supported Reagents : Preparation, Analysis and Applications, VCH*, New York, 1992.
[134] Ando, J. M. Miller and M. J. Robertson, *J. Chem. Soc., Perkin. Trans.*, 1986, 2, 1133.
[135] D. C. Neckers, D. A. Kooistra and G. W. Greeen, *J. Am. Chem. Soc.*, 1972, 9284.
[136] V. L. Magnotta, B. C. Gates and G. C. A. Schuit, *J. Chem. Soc. Chem. Commun.*, 1976, 342.
[137] V. L. Magnotta and B. C. Gates, *J. Pol. Sci. Polym. Chem.* Ed., 1977, 15, 1341.
[138] L. D. Rollmann, *J. Am. Chem. Soc.*, 1975, 97, 2132.
[139] M. Cinouini, S. Colonna and H. Molinari and F. Montanari, *J. Chem. Soc. Chem. Commun.*, 1976, 396.
[140] M. Anpo, *Pure Appl. Chem.*, 2000, 72, 1265.
[141] Takeuchi, Masato, H. Yamashita, M. Matsuoka, M. Anpo, T. Hirao, N. Itoh and N. Iwamoto, *Catal. Lett.*, 2000, 67, 135
[142] G. Colon, J. N. S. Espana, M. C. Hidalgo and J. A. Navio. *J. Photochem Photobiol*, 2006, 179A, 20.
[143] Nakajima, S. Koizumi, T. Watanabe and K. Hashimoto, *Langmuir*, 2008, 16, 7048.

[144] Hager S, Baver R, Kudielba G, *Photocatalytic oxidation of gaseous chlorinated organics over titanium dioxide. Chemosphere*, 2000, 41, 1219.
[145] Hager S, Baver R, *Heterogeneous photocatalytic oxidation of organics for air purification by near uv irradiated titianium dioxide. Chemosphere* 1999, 38, 1549.
[146] J. S. Yadav and H. M. Meshram, *Pure Appl. Chem*., 2001, 73, 199.
[147] G. W. Gribbe, *Chem. Soc. Rev*., 1999, 28, 335.
[148] M. K. Chaudhari, A. T. Khan, B. K. Patel, D. Dey, W. Kharmawphlang, T. R. Lakshmiprabha and G. C. Mandel, *Tetrahedron Lett*., 1998, 39, 8163.
[149] J. A. Cusumano, *Chemtech.*, 1992, 485.
[150] R. S. Oakes, A. A. Clifford, K. D. Bartle, M. Pett and C. M. Rayner, *Chem. Commun*., 1999, 247.
[151] D. L. Sparks, L. A. Estevez and R. Hernandez, *Green Chem*., 2009, 11, 986.
[152] R. Mello, A. Ohmos, J. Parra-Corbonell, M. E. Gonzalez-Nunez and G. Asensio, *Green Chem*., 2009, 11, 994.
[153] S. K. Karmee, C. Roosen, C. Kohlmann, S. Lutz, L. Greener and W. Leitner, *Green Chem*., 2009, 11, 1052.
[154] J. Wang, L. Hu and C. Miao, *Green Chem*., 2009, 11, 1013.
[155] R. S. Oakes, A. A. Clifford and C. M. Rayner, *J. Chem. Soc., Perkin Trans,* 2001, 1, 917.
[156] S. C. Ameta, *J. Indian Chem. Soc*., 2002, 79, 305.

In: Handbook of Research on Chemoinformatics ... ISBN: 978-1-62100-998-6
Editor: A.K. Haghi

Chapter 5

ADVANCES IN DOWNDRAFT BIOMASS GASIFICATION

Pratik N. Sheth and B. V. Babu*
Chemical Engineering Group, Birla Institute of Technology and Science (BITS), Rajasthan, India

1. ENERGY RESOURCES

All living organisms rely on an external source of energy to grow and to reproduce. The principal perpetual energy resources are solar energy, wind power and bioenergy, all of which ultimately depend on an extra-terrestrial source, namely the Sun. Most of the world's energy resources are from the sun's rays hitting earth – some of that energy has been preserved as fossil energy, some is directly or indirectly usable, e.g., via wind, hydro or wave power, some is interspersed in the biomass as a chemical energy via photosynthesis process. Prehistoric plants stored the Sun's energy in their leaves, branches and roots, and when they died and eventually converted to fossil fuel, the energy from which releases upon burning. Sun warms our atmosphere and the resulting heated air tends to rise and forms winds and waves. These energy resources may be categorized into two groups: finite (e.g. minerals) and perpetual (renewable resources such as solar, wind, tidal, etc.). Whilst each major energy source has its own characteristics, applications, advantages and disadvantages, the fundamental distinction among these is between those that are finite and those that are, on any human scale, effectively perpetual or everlasting. The finite resources comprise a number of organically-based substances such as coal, crude oil, oil shale, natural bitumen & extra-heavy oil, and natural gas, together with the metallic elements such as uranium and thorium. And the principal perpetual resources are solar energy, wind energy and bioenergy. Other perpetual resources are various forms of marine energy – tidal energy, wave energy and ocean thermal energy conversion (OTEC). There are other energy resource such as peat and geothermal energy, which are, to some extent intermediate in nature, with both finite and perpetual elements in their make-up. Bioenergy is arguably the one truly renewable energy resource, in that each new crop or harvest represents a partial renewal of its resource base, which itself is subject to constant depletion through its use as a fuel or feedstock. All the remaining

* Email: bvbabu@bits-pilani.ac.in

perpetual energy resources are available on a continuing, although varying basis, are not depleted by the utilization of their energy content, and are therefore not subject to renewal (Caillé, 2007).

The world's primary energy consumption is about 400 EJ/year as per estimates reported by Fridleifsson (2003) and is mostly provided by fossil fuels (80%). The effects on global and environmental air quality of pollutants released into the atmosphere from fossil fuels provide strong arguments for the substitution of these fossil fuels with renewable energy resources. Clean, domestic and renewable energy is commonly accepted as the key for future life. The renewable sources collectively provide 14% of the primary energy, in the form of traditional biomass (10%), large (>10 MW capacity) hydropower stations (2%), and the "new renewable sources" (2%). Nuclear energy constitutes 6% of the total energy. The World Energy Council (London, UK) expects the world primary energy consumption to have grown by 50–275% in 2050, depending on different scenarios. The renewable energy sources are expected to provide 20–40% of the primary energy in 2050 and 30–80% in 2100. The technical potential of the renewable sources is estimated at 7600 EJ/year, and thus certainly sufficiently large to meet future world energy requirements (Fridleifsson, 2003).

2. BIOENERGY

Biomass is a term used to describe all biologically produced matter and it is the name given to all earth's living matter. It is a general term for material derived from growing plants or from animal manure (which is effectively a processed form of plant material). The chemical energy contained in the biomass is derived from solar energy using the process of photosynthesis. This is the process by which plants take in carbon dioxide and water from their surroundings and, using the energy from sunlight, convert them into sugars, starches, cellulose, lignin, etc. which make up vegetable matter loosely termed carbohydrates (and shown for simplicity as $[CH_2O]$) and oxygen.

$$CO_2 + 2H_2O \longrightarrow [CH_2O] + H_2O + O_2 \tag{1}$$

Biomass energy is derived from the plant sources, such as wood from natural forests, waste from agricultural and forestry processes and industrial, human or animal wastes. The stored energy in the plants and animals (that eat the plants and other animals), or the waste that they produce is called biomass energy or bioenergy. It is a natural process that the entire biomass ultimately decomposes to its molecules with the release of heat. And the combustion of biomass imitates the natural process. So the energy obtained from biomass is a form of renewable energy and it does not add carbon dioxide to the environment in contrast to the fossil fuels (Twidell, 1998). Of all the renewable energy sources, biomass is unique in that it effectively stores solar energy inherently. Furthermore, it is the only renewable energy source of carbon and is able to convert into convenient solid, liquid and gaseous fuels (Demirbas, 2001).

Bioenergy is essentially renewable or carbon neutral. Carbon dioxide released during the energy conversion of biomass (such as combustion, gasification, pyrolysis, anaerobic digestion or fermentation) circulates through the biosphere, and is reabsorbed in equivalent stores of biomass through photosynthesis. Fig. 1 shows the combustion of wood and thereby CO_2 generation. It also depicts that net CO_2 generation is zero as new biomass could be developed photosynthetically for the growth of which this CO_2 is utilized. Biomass energy is a renewable and unique form of solar energy. Of the massive 178,000 x 10^{12} Watts of solar energy that falls on the Earth's surface, some 0.02% or around 40 x 10^{12} Watts is captured by plants via photosynthesis and is available as biomass energy. This translates into the production of some 220 billion 'dry' tonnes of biomass per year, which as an energy source represents some ten times the world's total current energy use. Currently some 15 percent of the planet's energy requirements are met from biomass, mainly for cooking and heating in developing countries. Biomass is also supplied as a fuel in a growing number of large scale, modern biomass energy plants in industrialized countries. By comparison, the world population consumes around 10 EJ/year of energy in the form of food, which of course is a biomass energy resource in itself (Stucley et al., 2004).

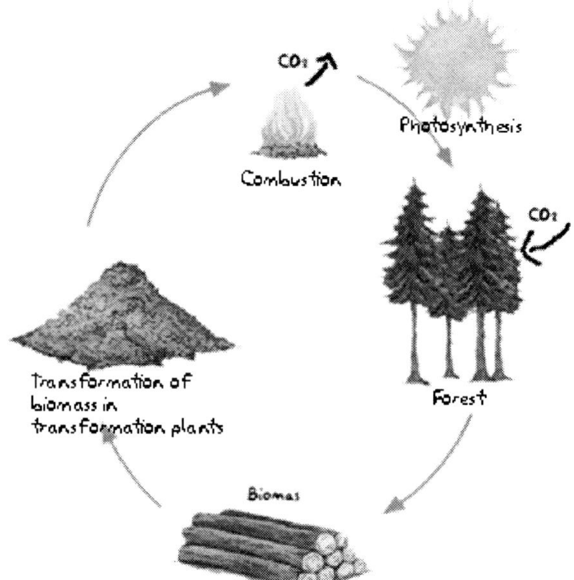

(Source: http://www.paisatge.net/SapreRenovables/ENG/eSproj.htm)

Figure 1. Lifecycle of forest biomass.

3. Biomass Conversion Technologies

Conventionally biomass was used in a similar way to fossil fuels, by burning it at a constant rate in a boiler furnace to heat water for producing steam. This steam passes through the multiple blades of turbine, spinning the shaft. The turbine shaft drives an electricity generator which produces an alternating current for local use or to supply the national grid. Wood is still a predominant fuel in many non-OPEC (Organization of the Petroleum Exporting Countries), tropical, developing countries and it will continue to be used for many years. It competes well with fossil fuels, because it is renewable, and with soft energies (solar and wind), on account of its energy storage capability. It is being used in the domestic sector (for cooking and water heating), commercial sector (water heating) and industrial sector (for water heating and process heat); and also in rural industries, such as brick kilns, potteries etc. (Demirbas, 2000). In nature, biomass is not concentrated, and so, the use of naturally occurring biomass requires transportation, which increases the cost and reduces the net energy production. Biomass has a low bulk density, which makes transportation and handling more difficult and costly. Apart from transportation, incomplete combustion of biomass generates a concern among the environmentalists, as it may produce organic particulate matter, carbon monoxide and other organic gases. If high temperature combustion is used, oxides of nitrogen would be produced. The health impact of air pollution is a significant problem in developing countries, where fuel wood is burnt inefficiently in open fires for domestic cooking and space heating (Demirbas, 2001). The conversion technologies for utilizing biomass can be separated into three basic categories: direct combustion processes, thermochemical processes and biochemical processes. Direct combustion deals mainly with primary fuels, e.g., the form in which it is available in nature or after some form of processing (drying, sizing, briquetting, etc.). In the other two processes the primary fuel is converted into a secondary fuel (solid, gas and/or liquid form) by processes such as pyrolysis, gasification, carbonization, digestion, fermentation, etc. The secondary fuels obtained from the conversion process can be used directly for various end-use activities for further processing (Babu, 2008a).

Direct combustion is used to convert biomass into a useful energy. The heat and/or steam produced during this process are/is used to provide process heat for domestic cooking and industrial processes or to generate electricity. Biomass-fired power plants have been installed in a number of countries in Asia. The main disadvantage of this process is the incomplete and inconsistent combustion of solid fuel. Direct combustion of biomass materials usually results in smoke and ash pollution unless special filtering equipment is used. Biochemical processes make use of the biochemistry of the raw materials, and the action of microbial organisms, to produce gaseous and liquid fuels like biogas, ethanol and methanol. Digestion is the biochemical conversion of organic material to biogas, which mainly consists of methane and carbon dioxide. Anaerobic reactors are generally used for the production of biogas from manure and crop residues. Fermentation is another example of biochemical process in which micro-organisms (usually yeast) break down sugars to produce ethanol. Ethanol is produced from certain biomass materials that contain sugars, starch or cellulose. It is regarded as an important potential alternative source of liquid fuels for the transport sector. Ethanol blended with conventional fuel like petrol or diesel is widely used now-a-days. Although fermentation is successfully used to produce ethanol, it is feedstock limited, time consuming and low yield process. There has been an increasing interest for thermochemical conversion of biomass and urban wastes for upgrading the energy in terms of more easily handled fuels, namely gases, liquids, and charcoal in the past decade. The thermochemical conversion of biomass (pyrolysis, gasification, combustion) is one of the promising routes amongst the renewable energy options of future energy. It is renewable form with many ecological advantages (Babu, 2008a).

Thermochemical conversion processes can be subdivided into gasification, pyrolysis, and direct liquefaction. Gasification is a process of conversion of solid carbonaceous fuel into combustible gas by partial combustion. The resulting gas, known as producer gas, is a mixture of carbon monoxide, hydrogen, methane, carbon dioxide and nitrogen. The producer gas is more versatile than the original solid biomass. It is burnt to produce process heat and steam or used in gas turbines to produce electricity. In both pyrolysis and liquefaction processes, feed stock organic compounds are converted into liquid products. In case of liquefaction, the macromolecules of feedstock compounds are decomposed into fragments of light molecules in the presence of a suitable catalyst. These unstable and highly reactive fragments repolymerize into heavy compounds having high molecular weights. Whereas in pyrolysis, catalyst is not used and light decomposed fragments are converted to oily compounds through homogeneous reactions in the gas phase. The difference in operating conditions for liquefaction and pyrolysis are shown in Table-1 (Demirbas, 2000).

Table 1. Comparison of liquefaction and pyrolysis

Process	Temperature (°C)	Pressure (MPa)	Drying
Liquefaction	525-600	5-20	Unnecessary
Pyrolysis	650-800	0.1-0.5	Necessary

In the pyrolysis process, biomass gets decomposed by heat in the absence of oxygen, which results in the production of various organic gaseous products, charcoal and tar. The study of pyrolysis is attracting researchers, as it is not only an independent process, but also a first step in the gasification or the combustion process (Babu and Chaurasia, 2003a-b).

4. Biomass Gasification

Gasification was discovered independently in both France and England in 1798. By 1850 the technology had been developed to the point that it was possible to light much of London with the manufactured gas or "town gas" from coal (Singer, 1958). During World War I and II wood gas generators, called Gasogene, were used when fuel supply was not enough but after a few years, a more reliable and cheap technology was developed that runs on petroleum; and gradually the use of gas produced by biomass was reduced. Due to the energy crisis in most of the countries and towering cost of petroleum, biomass based gasification process is again in focus in the recent past (Reed and Das, 1988). Gasification of biomass is one of the majorly used processes to increase the efficiency of energy harnessing from biomass. Gasification is a process that takes carbonaceous materials as its feed, such as coal, petroleum, or biomass, and converts into carbon monoxide and hydrogen. The raw material reacts with a controlled amount of oxygen and/or steam at high temperatures (Reed and Das, 1988). It is also a very efficient method for extracting energy from many different types of organic materials, and also has applications as a clean waste disposal technique. Moreover the usage of producer gas is potentially more efficient than direct combustion of the original fuel because it can be combusted at higher temperatures. The typical composition of hydrogen in producer gas varies from 5-25% depending upon moisture content of the fuel. After separation and purification, it can be utilized in fuel cell and biomass gasification process can be considered as a one of the prominent process for biohyrogen production (Babu, 2008b). There are mainly two techniques available for gasification of biomass, viz., fixed bed mode and fluidized bed mode (Bhave, 2001). The three main configurations of fixed bed gasifiers include Updraft, Downdraft, and Crossdraft mode of operations. In the updraft gasifiers, biomass moves down vertically and comes in contact with an upward moving product gas stream counter-currently. The updraft gasifier is easy to build and operate but product gas is very dirty with high amount of tar. It also has a high thermal efficiency as gases from the combustion zone passes upwards through incoming fuel, which preheat it (Bridgwater, 2002). In the downdraft gasifier biomass moves slowly downwards and air is introduced cocurrently and reacts at a throat that supports the gasifying biomass. They are cheap and easy to make. A relatively clean gas is produced with low tar and usually with high carbon conversion. In the cross draft gasifier, air is introduced on one side of the gasifier and the gas outlet is on the opposite side. Normally an air inlet nozzle is extended to the center of the combustion zone. The main advantages of the crossdraft gasifiers are: (1) its rapid response to change in load, (2) its simple construction and (3) its lightweight. Cross draft gasifiers are best suited for clean fuel like charcoal (Reed and Das, 1988). Out of different configuration of reactors for biomass gasification, a survey of gasifier manufacturers have reported that 75% of gasifiers offered commercially were downdraft, 20% were fluid beds [including circulation fluid beds], 2.5% were updraft, and 2.5% were of other types (Bridgwater, 2002).

Fig. 2 is a schematic diagram of the down draft biomass gasifier. The nozzle and constricted hearth downdraft gasifier is sometimes called an "Imbert" gasifier, named after its entrepreneurial inventor, Jacques Imbert (Reed and Das, 1988).

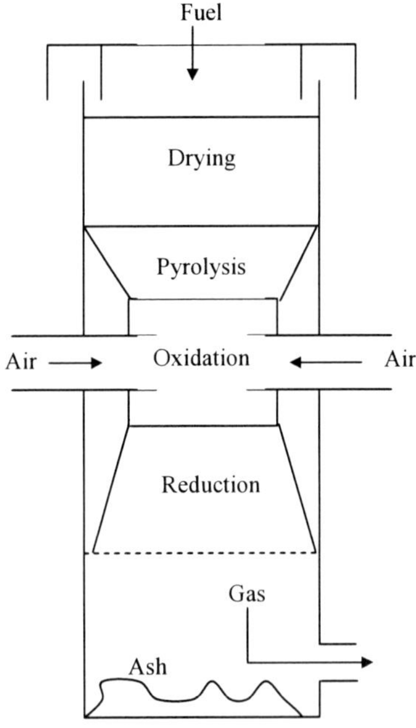

Figure 2. Schematic diagram of an Imbert downdraft biomass gasifier.

It may be divided into four distinct zones: (1) Drying/Preheating zone (2) Pyrolysis/Devolatilization zone (3) Combustion zone and (4) Reduction/Gasification zone. The description of these four zones is given in the subsequent sections.

4.1 Drying

The first zone in which the feed comes in contact in biomass gasifier is the drying zone. Basically drying is a mass transfer operation resulting in the removal of water moisture by evaporation from a solid or semi-solid. To achieve this, there must be a source of heat, and a sink of the vapor thus produced. This process should ideally take place at a temperature of around 160°C using waste heat from the conversion process. In the drying zone, feed descend into the gasifier and moisture is removed using the heat generated in the zones below by evaporation. The water vapor flows downward in the gasifier. Part of it may be reduced to hydrogen in the reduction zone and the rest will end up as moisture in the gas. The rate of drying depends on the surface area of the fuel, the temperature difference between the feed and the hot gases, the re-circulation velocity, relative humidity of these gases, and the internal diffusivity of moisture within the fuel (Dogru et al., 2002). In this zone, no chemical reaction takes place and only the water removal is carried out.

4.2 Pyrolysis

The pyrolysis of biomass is a promising route for the production of solid (charcoal), liquid (tar and other organics such as acetic acid, acetone and methanol) and gaseous products (H_2, CO_2, CO). These products are of interest as they are possible alternate sources of energy. Pyrolysis is a process by which a biomass feedstock is thermally degraded in the absence of oxygen/air. The basic phenomena that take place during pyrolysis are: (1) heat transfer from a heat source, leading to an increase in temperature inside the fuel; (2) initiation of pyrolysis reactions due to this increased temperature, leading to the release of volatiles and the formation of char; (3) outflow of volatiles, resulting in heat transfer between the hot volatiles and cooler unpyrolysed fuel; (4) condensation of some of the volatiles in the cooler parts of the fuel to produce tar; and (5) autocatalytic secondary pyrolysis reactions due to these interactions (Babu, 2008a; Babu and Chaurasia, 2004a-d; Chaurasia and Babu, 2004). Pyrolysis can also be used as an independent process for the production of useful energy (fuels) and/or chemicals. The overall process of pyrolysis can be classified into primary and secondary stages. When a solid particle of biomass is heated in an inert atmosphere the following phenomena occur. Heat is first transferred to the particle surface by radiation and/or convection and then to the inside of the particle. The temperature inside the particle increases, causing (1) removal of moisture that is present in the biomass particle, (2) the pyrolysis reactions to occur. The heat changes due to the chemical reactions, and phase changes contribute to a temperature gradient as a function of time, which is nonlinear. Volatiles and gaseous products flow through the pores of the particle and participate in the heat transfer process. The pyrolysis reactions proceed with a rate depending upon the local temperature. During the pyrolysis process, the pores of the solid are enlarged, and the solid particle merely becomes more porous because the biomass converts into gases (Curtis and Miller, 1988). According to Anthony and Howard (1976), the enlarged pores of the pyrolyzing solid offer many reaction sites to the volatile and gaseous products of pyrolysis and favor their interaction with the hot solid. Inside the pyrolyzing particle, heat is transmitted by the following mechanisms: (a) conduction inside the solid particle; (b) convection inside the particle pores; and (c) convection and radiation from the surface of the pellet.

Depending upon the operating conditions, the pyrolysis process can be divided into three subclasses: conventional pyrolysis (carbonization), fast pyrolysis, and flash pyrolysis. The ranges of the main operating parameters for pyrolysis processes are given in Table-2. Conventional pyrolysis is defined as the pyrolysis that occurs under a slow heating rate. This condition permits the production of solid, liquid, and gaseous pyrolysis products in significant portions. The first stage of biomass decomposition occurs between 395 and 475 K and is called pre-pyrolysis. During this stage, some internal rearrangement such as water elimination, bond breakage, appearance of free radicals, and formation of carbonyl, carboxyl, and hydroperoxide groups takes place (Shafizadeh, 1982). The second stage of solid decomposition corresponds to the main pyrolysis process. It proceeds with a high rate and leads to the formation of the pyrolysis products. During the third stage, the char decomposes at a very slow rate and carbon-rich residual solid forms. If the objective is to the produce mainly liquid and/or gaseous products, a fast pyrolysis is recommended. The achievement of fast heating rates requires high operating temperatures, very short contact times, and very fine particles. Flash pyrolysis differs strongly from that of conventional pyrolysis, which is performed slowly with relatively massive pieces of wood. Flash pyrolysis gives mostly gaseous products due to high heating rate and very small particle size (Demirbas and Arin, 2002). Hydropyrolysis (pyrolysis in a hydrogen atmosphere) is also considered to have a potential application in the conversion of biomass to liquids enriched in hydrocarbons (Barth, 1999). Most biomass materials are chemically and physically heterogeneous, and their components have different reactivities and yield different products. Biomass is mainly composed of three constituents which are hemicellulose, cellulose, and lignin. It also contains minor amounts of extractives. Each component of biomass pyrolyzes at different rates and by different mechanisms and pathways.

Table 2. Ranges of main operating parameters for pyrolysis processes

Parameters	Conventional Pyrolysis	Fast Pyrolysis	Flash Pyrolysis
Pyrolysis Temperature (K)	550-950	850-1250	1050-1300
Heating rate (K/s)	0.1-1.0	10-200	>1000
Particle size (mm)	5-50	<1	<0.2
Solid residence time (s)	450-550	0.5 -10	<0.5

It is believed that as reaction progresses the carbon becomes less reactive and forms stable chemical structures, and consequently the activation energy increases as the conversion level of biomass increases. Cellulose and hemicellulose decompose over a very narrow temperature range as compared to lignin. The rate and extent of degradation of each of these components depend on the process parameters such as reactor type, temperature, particle size, heating rates and pressure (Bridgwater et al., 1999). The hemicellulose breaks down first, at a temperature of 470 to 530 K, and cellulose follows in the temperature range 510 to 620 K, with lignin being the last component to pyrolyze at temperatures of 550 to 770 K (Demirbas and Küçük, 1999).

4.3 Combustion

In the oxidation or combustion zone, biomass along with the volatile products of pyrolysis are oxidized resulting in a rapid rise in temperature up to 1200°C due to highly exothermic reactions. The heat generated is used to drive the drying and pyrolysis of the fuel and the gasification reactions. The oxidation reactions of volatiles are very rapid and the oxygen is consumed before it can diffuse to the surface of the char. Therefore, no combustion of solid char can take place. Oxidation of the condensable organic fraction to form lower molecular weight products is important in reducing the amount of tar produced by a gasifier (Dogru et al., 2002). Biomass combustion is more complex than either pyrolysis or gasification since the biomass must first pyrolysed, then be partially combusted (gasified) before it is fully combusted. However, the overall global reaction of biomass oxidation can be represented by Eq. (1.2) (Reed and Das, 1988).

$$CH_{1.4}O_{0.6} + 1.05O_2 + 3.95N_2 \longrightarrow CO_2 + 0.7H_2O + 3.95N_2 \qquad (2)$$

where $CH_{1.4}O_{0.6}$ is an average formula for woody biomass.

4.4 Reduction

The gaseous mixture leaving the combustion zone mainly containing carbon dioxide, water vapor, inert nitrogen, and some amount of low molecular weight hydrocarbons such as methane, ethane, ethylene etc., passes over the hot charcoal in the reduction zone, which is often referred as gasification zone. The principal reaction in the reduction zone is that of carbon dioxide with hot carbon to produce carbon monoxide. This is an endothermic process. It is referred to as the 'Boudouard' reaction (Eq. (1.3)).

$$C + CO_2 \longrightarrow 2CO \ \Delta H = 1{,}72{,}600 \text{ J/mol} \qquad (3)$$

Another important reaction occurs between water vapor and carbon resulting in the formation of carbon monoxide and hydrogen as given by Eq. (1.4).

$$C + H_2O \longrightarrow CO + H_2 \ \Delta H = 1{,}31{,}400 \text{ J/mol} \qquad (4)$$

Reaction (1.4) is called the water gas reaction. This is also an endothermic reaction and takes place between 600°C and 950°C. As the reactions (1.3) and (1.4) are endothermic, gas stream loses heat and the temperature drops in the reduction zone progressively. If excess water is present in the reduction zone the so-called water shift reaction can also takes place (Reaction (1.5)).

$$CO + H_2O \longrightarrow CO_2 + H_2 \ \Delta H = -\ 41{,}200 \text{ J/mol} \qquad (5)$$

This is an exothermic reaction and is undesirable as it reduces the calorific value of gas. So the excess moisture in the fuel is to be avoided. Most of the hydrogen that is produced in the reduction zone remains free. However, a portion of it can combine with carbon to form small amounts (about 3 to 5%) of methane as given by reaction (1.6).

$$C + 2H_2 \longrightarrow CH_4 \; \Delta H = -75{,}000 \text{ J/mol} \quad (6)$$

The gaseous mixture leaving the biomass gasifier mainly contains carbon monoxide, hydrogen, carbon dioxide, nitrogen and water vapor. It may also contain some amount of hydrocarbons such as CH_4, C_2H_2 and C_2H_6, the amount of each may depend upon the configuration of the gasifier. Producer gas is also loaded with dust, tar and water vapor.

5. Experimental Studies on Downdraft Biomass Gasification

Various group of researchers carried out experimental study on downdraft biomass gasifier with different biomass materials. Sheng (1989) introduced the properties of biomass gasifier and the gasifying reaction processes. Various fuels usually considered unusable or low-value wastes, such as sawdust, wood chips, corn cobs, nut shells, rice hulls, etc, were used in the biomass gasifier to produce a high calorific value gas. It was found that gas could replace conventional energy sources, such as fuel oil, electricity and natural gas for crop drying, space heating, and industrial boilers. It was concluded that the gas produced by downdraft gasifier is tar-free and could be used directly to drive most of the internal combustion engines. Rice husk was successfully used as a biomass material in a downdraft biomass gasifier by Chowdhury et al. (1994). In their experimental study, the air velocity was varied in the range of 0.032-0.099 m/s, and gasification rate was found to be in the range 1.8×10^{-2} - 4.3×10^{-2} kg/m^2s. The calorific value of producer gas obtained from the gasifier was found in the range of 3240-4382 kJ/m^3. Jorapur and Rajvanshi (1995) developed a gasifier running on chopped sugarcane leaves (1 – 10 cm). The experimental system was consisted of a throatless cylindrical gasifier, a gas conditioning system and a diesel-powered generator along with its control panel. The gas conditioning system consisted of a cyclone, an impact filter, an indirect-contact heat exchanger, a centrifugal scrubber and a bubbling-cum-packed bed filter. A 15 kVA diesel generator set was operated for over 200 h using the gasifier. The gas flow rate was varied in the range of 3-4 Nm^3/hr. The gas compositions, calorific values and cold gas efficiency were found. The calorific value of the producer gas obtained was 3.5 to 5.0 MJ /Nm^3. The cold gas efficiency found to be 35-60% over the entire range of loads tested (3.5-11.3 kW). About 15-28% by weight of the fuel was converted into char with a calorific value of 19 MJ kg^{-1}. Jorapur and Rajvanshi (1997) in their subsequent study reported the development of a commercial-scale (1080 MJ/h) gasification system using low-density biomass such as sugarcane leaves, bajra stalks, sweet sorghum stalks and bagasse for thermal applications. The system was tested for higher than 700 h of operations under laboratory conditions at 288–1080 MJ/h output levels. The calorific value of the gas was reported as 3.56–4.82 MJ/m^3. After successful laboratory trial, the system was tested at a metallurgical company, where it was retrofitted to an oil-fired furnace for baking speciality ceramics. Di Blasi et al (1999) designed a laboratory scale countercurrent fixed-bed gasification plant and constructed to produce data for process modeling and to compare the gasification characteristics of several biomass materials such as beechwood, nutshells, olive husks, and grape residues. The composition of producer gas and spatial temperature profiles were measured for biomass gasification at different air flow rates. It was found that the gas heating value reached a maximum with a variation of air-to-fuel ratio. The gas heating values were in the range of 5 – 5.5 MJ/Nm^3 with 28-30% CO, 5-7% CO_2, 6-8% H_2, 1-2% CH_4, and small amounts of C_2-hydrocarbons (apart from nitrogen). It was reported that the gasification of agricultural residues was more difficult because of bed transport, partial ash sintering, non uniform flow distribution and the presence of muddy phase in the effluents, which might require proper pretreatments.

Zainal et al. (2002) carried out an experimental investigation of a downdraft biomass gasifier using furniture wood and wood chips. The effect of equivalence ratio on gas composition, calorific value and the gas production rate was presented. It was found that the calorific value of producer gas increased with equivalence ratio initially, attained a peak and then decreased with an increase in equivalence ratio. The gas flow rate per unit weight of the fuel was found increasing linearly with equivalence ratio. It was also observed that the complete conversion of carbon to gaseous fuel did not occur even for the optimum equivalence ratio. Dogru et al. (2002) used a pilot scale downdraft gasifier to investigate the gasification potential of hazelnut shells. A full mass balance was used including the tar production rate. The effect of feed rate on composition of product gas and the gasifier zone temperatures were determined. The optimum value of air-to-fuel ratio found to be ranging from 1.44 to 1.47 Nm^3/kg. At the optimum air-fuel-ratio, the volumetric flow rate of the producer gas was found to be 8.5 Nm^3/h with a calorific value of about 5 MJ/m^3. It was reported that the hazelnut shells could be easily gasified in a downdraft gasifier to produce good quality gas with a minimum of polluting by-products. It was suggested that, in view of the ease of operation, small-scale gasifier could make an important contribution to the economy of rural areas where the residues of nuts were abundantly available. Jayah et al. (2003) carried out experimental studies on a gasifier, which was fabricated in Sri Lanka for tea industry, to examine the effect of certain key operating parameters and design features on its performance. It was found that wood chips of size 3–5 cm with moisture content below 15% (dry basis) should be used in this gasifier. Wander et al. (2004) analyzed the sawmill residues as an energy source to offer the best technical, economic and environmental alternative. The characterization (quantity, type, chemical and energetic analysis) of the residues generated, in addition to the energy needs of sawmills, was supportive to find the suitable technology. It was found that the technology of wood gasification could produce a gas that is able to burn in an internal combustion engine after appropriate cleaning. In order to assess the performance of the wood residues gasification process, a small, fixed bed, downdraft, stratified and open top gasifier was built. The gasifier capacity was around 12 kg/h. An internal gas recirculation was introduced, in which part of the produced gas was burnt to raise the gasification reaction temperature. The mass and energy balances of the gasifier were carried out and its cold gas, global and mass conversion efficiencies were found out using several parameters measured in the experiments. Sharma (2007) used an open top downdraft biomass gasifier of capacity 20 kWe to validate the model developed in his earlier studies (Sharma, 2006). Steady state temperatures at different locations inside the reactor were measured using K – type thermocouples at various values of pressure and air flow rate. Sheth and Babu (2009) carried out an experimental study on a downdraft biomass gasifier with the waste generated while making furniture in the carpentry section of the workshop of their organization. *Dalbergia sisoo*, generally known as sesame wood or rose wood is mainly used in the furniture and wastage of the same was used as a biomass material in their gasification studies. The effects of air flow rate and moisture content on biomass consumption rate and quality of the producer gas generated were studied by performing experiments. The performance of the biomass gasifier system was evaluated in terms of equivalence ratio, producer gas composition, calorific value of the producer gas, gas production rate, zone temperatures and cold gas efficiency. Material balance was carried out to examine the reliability of the results generated. The experimental work carried out by various researchers is summarized in Table-3. Equivalence ratio is defined as given by Eq. (7)

$$\Phi = \frac{\left(\text{Air Flow Rate} / \text{Biomass Consumption Rate} \right)}{\left(\text{Air Flow Rate} / \text{Biomass Consumption Rate} \right)\Big|_{\text{Stoichiometric}}} \quad (7)$$

The stoichiometric ratio of air flow rate to the rate of biomass consumption is 5.22 m^3 air/kg of wood (Zainal et al., 2002).

It can be seen from Table-3 that various kinds of biomass materials are used in the gasification studies. Each experimental study is carried out with different perspective. Few researchers such as Sheng (1989), Chaudhary et al. (1994), Jorapur and Rajvanshi (1995; 1997) carried out the experiments primarily to show how well biomass gasification can be applied to various agricultural wastes. Hence, they have not reported the air flow rate or biomass consumption and in turn equivalence ratio can not be found. They have also not suggested the optimum operating conditions. The other researchers such as Di Blasi et al (1999), Zainal et al. (2002), Dogru et al. (2002), Jayah et al. (2003), Wander et al. (2004), Sharma (2007) and Sheth and Babu (2009) carried out the gasification study to validate the model they have proposed. Out of these experimental research publications, Dogru et al. (2002) and Sheth and Babu (2009) reported their experimental findings in the most comprehensive way. However, Dogru et al. (2002) carried out experimental study with very narrow range of equivalence ratio. Hence, in this chapter the experiments carried out by Sheth and Babu (2009) is incorporated, which is the recent most addition in the scientific literature of biomass gasification experimental study

Table 3. Summary of the experimental work carried out by various researchers

Reference	Biomass used	Operating conditions studied		Equivalence Ratio (Φ)	Temperature measurements carried out	Optimum operating conditions suggested	Calorific Value of Gas (MJ/Nm^3)
		Biomass consumption (kg/h)	Air Flow rate (m^3/hr)				
Sheng (1989)	sawdust, wood chips, corn cobs, nut shells, rice hulls	25-150	not reported	not reported	Y	N	4.67
Chowdhury et al. (1994)	rice husk	not reported	3.6-11.3	not reported	N	N	3.2-4.4
Jorapur and Rajvanshi (1995)	chopped sugarcane leaves (1 – 10 cm)	1.2-3.5	not reported	not reported	Y	N	3.5-5.0
Jorapur and Rajvanshi (1997)	sugarcane leaves, bajra stalks, sweet sorghum stalks and bagasse	not reported	not reported	not reported	Y	N	3.5-5.0
Di Blasi et al. (1999)	beechwood, nutshells, olive husks, and grape residues	0.99-2.21	1.0-2.83	0.19-0.25	Y	N	5.0-5.5
Zainal et al. (2002)	wood chips	not reported	not reported	0.27-0.43	Y	N	4.65-5.62
Dogru et al. (2002)	hazelnut shells	1.73-5.4	2.82-8.1	0.26-0.31	Y	Y	3.47-5.12
Jayah et al. (2003)	wood chips	not reported	not reported	0.36-0.45	Y	N	not reported
Wander et al. (2004)	sawmill residues	12.0	not reported	0.19-0.55	Y	N	4.1-6.32
Sharma (2007)	wood blocks	not reported	not reported	0.44	Y	N	not reported
Sheth and Babu (2009)	wood waste	1.0-3.63	1.85-3.39	0.17-0.4	Y	Y	4.5-6.34

6. Experimental Setup and Procedure

6.1 Experimental Setup

An Imbert downdraft biomass gasifier is shown in Fig.3a. In downdraft gasifiers, pyrolysed gas and moisture generated in pyrolysis and drying zone respectively flow downwards. The pyrolysis gases pass through a combustion zone followed by a hot bed of char which is supported by a grate. Biomass is fed to the gasifier and oxidized in the zone where continuous air is supplied from two air nozzles.

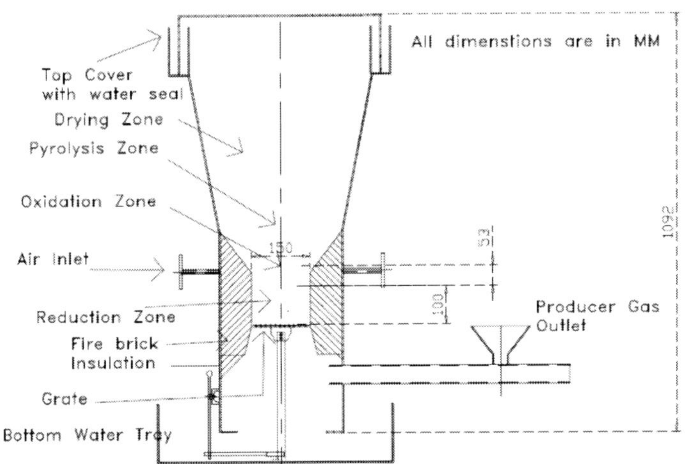

(a) Various parts, zones and dimensions of the unit

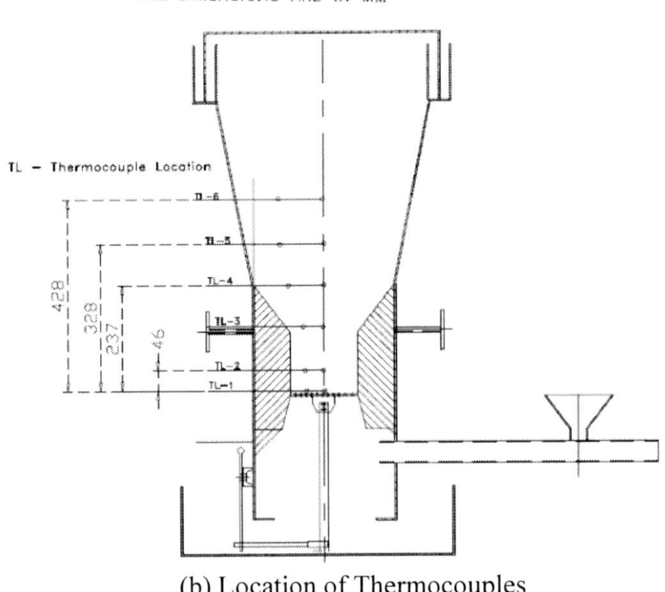

(b) Location of Thermocouples

Figure 3. Schematic Diagram of an Imbert downdraft Biomass Gasifier.

The heat generated in the combustion zone is transferred to the pyrolysis and drying zone. Released heat from the biomass combustion raises the temperature of the biomass particles resting above the oxidation zone and thus they get pyrolysed. The biomass particles are decomposed into volatiles and charcoal in the pyrolysis zone. Released volatiles from each of the biomass particles flow downward in the packed pyrolysis bed. The rate of volatiles release depends on particle size and temperature within the single particle. Due to high temperature of the combustion zone, tar of the pyrolysed gas mixture cracks into non-condensable gases and water. The cracked pyrolysed gas mixes with the carbon dioxide generated due to combustion and the inert N_2 present in the air. This gaseous mixture passes over the hot bed of charcoal and undergoes endothermic reduction reactions. Water seal is provided at the top of the gasifier to maintain the downward flow. The total height of the gasifier is 1.1 meter and the diameter at the pyrolysis zone is 310 mm and the diameter at the reduction zone is 150 mm. The height of the reduction zone is 100 mm and that of oxidation zone is approximately 53 mm. The height of the pyrolysis zone depends upon the biomass loading. The charcoal in the reduction zone is supported by a movable grate at the bottom of the gasifier. The ash produced during gasification is removed by rotating the grate using the lever arrangement provided to unclog the grate. By moving the grate, grate clogging and bridging of the biomass can be avoided thereby causing greater amount of biomass to undergo gasification.

To measure the temperature of various zones in the gasifier, thermocouples are placed inside the gasifier at different locations. There are six pairs of chromel – alumel thermocouples, each pair placed at different heights of the gasifier (Fig. 3b) so as to cover all the zones. In each pair, one thermocouple is placed at the center of the gasifier ($r = 0$) while the other is placed at a half radius distance ($r = R/2$). Two pairs of thermocouples are placed in the reduction zone of the gasifier, one at the end of the reduction zone that is close to the grate and one at a height of 46 mm from the grate. One pair of thermocouples is placed in the oxidation zone present at the level of the air inlet nozzles. Three pairs of the thermocouples are placed in the pyrolysis zone at height of 237 mm, 328 mm, and 428 mm from the grate respectively in the biomass gasifier.

6.2 Experimental Procedure

Water is filled in the container placed below the gasifier and also in the circular trough at the top of the gasifier which act as seals and hence to prevent the gas from escaping out of the gasifier. 500 grams of charcoal (collected from the residue of the wood based furnace, used for mass cooking locally) is dumped as a heap into the reduction zone of the gasifier above the grate. The initial charcoal used in the present experimentation is of the same quality as that generated in the gasification experiments by pyrolysis of wood. The air is introduced in the biomass gasifier through nozzles and its flow rate, which is maintained constant through a gate valve, is measured using a rotameter. Biomass is dumped in to the oxidation zone of the biomass gasifier and around 25 ml of diesel is poured to aid the combustion of biomass. Once combustion starts properly and spread across the oxidation zone which generally takes about 3 to 4 minutes, additional 3 kg of biomass is dumped inside the gasifier and it is closed from the top by cover. It is observed that by the time combustion starts properly in the oxidation zone most of the diesel gets combusted due to volatility of diesel and very high temperature generated in the oxidation zone. The time at which the cover is closed, is taken to be the starting time of the experiment. Temperatures are recorded at each location after every five minutes. The samples of producer gas coming out from the gas burner are collected in the syringes at five minutes interval. Results presented in this study are the averaged values obtained over 5 minutes interval each for five intervals. The averaging should take care of the variation with respect to time. Sampled gas is analyzed with gas chromatograph (NUCON 5765) with thermal conductivity detector. Each experimental run is carried out for 25 minutes. At the end of the experiment any leftover biomass and charcoal is removed from the gasifier.

7. Results and Discussion

7.1 Biomass Characteristics

The furniture waste of *dalbergia sisoo*, generally known as sesame wood or rose wood is used as a biomass in the present gasification studies. Table-4 lists the physical properties, the proximate analyses, ultimate analyses and chemical analyses of the *dalbergia sisoo*. Higher heating value is calculated using the empirical formula given by Eq. (1) reported in the literature (Parikh et al., 2005).

$$HHV\ (MJ/kg) = 0.3536\ FC + 0.1559\ VM - 0.0078\ ASH \tag{1}$$

Table 4. Characteristics of Dalbergia Sisoo

Physical Properties			
Size (mm^3)	Absolute Density (kg/m^3)	Bulk Density (kg/m^3)	
25.4 x 25.4 x 25.4	1170	605	
Proximate Analysis (% by wt. dry basis)*			
Fixed Carbon (FC)	Volatile Matter (VM)	ASH	Calculated HHV (MJ/kg)
15.70	80.40	3.90	18.06
Ultimate Analysis (% by wt. dry basis)			
Carbon	Hydrogen	Oxygen	Nitrogen
48.6	6.2	44.87	0.33
Chemical Analysis (% by wt.)			
Cellulose	Hemi cellulose	Lignin	Extractives
36.75	11.30	43.65	8.30

*Source : Bhave (2001)

7.2 Biomass Gasification

The details on the range of parameters varied in this experimental study are shown in Table-5. Biomass consumption rate is found to vary from 1.0 to 3.6 kg/hr for an air flow rate varying from 1.85 to 3.4 m^3/hr respectively. Moisture content is varied from 0.0437 to 0.1145 wt fraction on wet basis. The values of Equivalence ratio are calculated for each run and reported in Table-5.

Table 5. Biomass gasification experimental run details

Run	Air flow rate (m^3/hr)	Initial Moisture content (wt fraction, wet basis)	Biomass consumption Rate (kg/hr)	Equivalence Ratio(Φ)
1	2.7765	0.1145	2.10	0.2533
2	3.3935	0.0437	3.63	0.1791
3	1.8510	0.0437	2.12	0.1673
4	2.7765	0.0437	2.67	0.1992
5	2.7765	0.073	2.59	0.2054
6	1.8510	0.10	1.00	0.3546

7.3 Effect of Moisture Content

The effect of moisture content on biomass consumption rate is shown in Fig. 4. It is found that with an increase in the moisture content, the biomass consumption rate decreases. For higher moisture content of biomass, the energy requirement for drying increases and reduces the biomass pyrolysis. The biomass moisture content greatly effects both the operation of the gasifier and the quality of the product gas. The constraint of moisture content for gasifier fuels are dependent on type of gasifier used. Higher values of moisture content could be used in updraft systems but the upper limit acceptable for a downdraft reactor is generally considered to be around 40% on dry basis (Dogru et al., 2002).

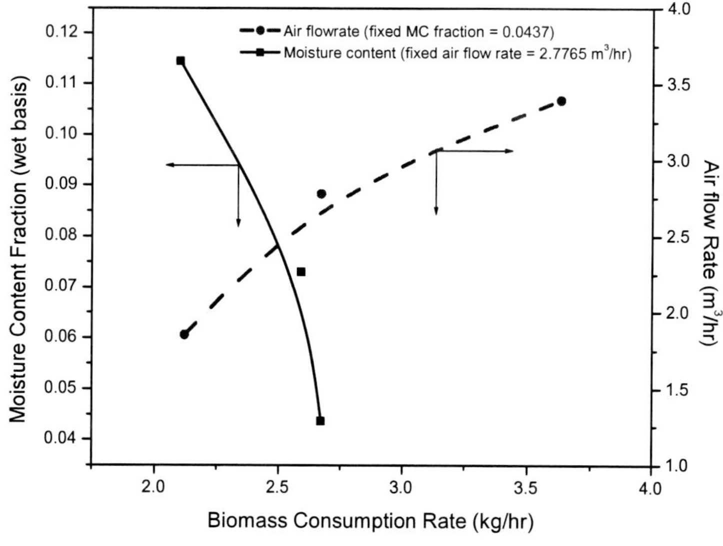

Figure 4. Effect of Moisture content (for an air flow rate of 2.7765 m^3/hr) and of Air flow rate (for a MC of 0.0437) on Biomass consumption rate.

7.4 Effect of Air Flow Rate

The effect of air flow rate on biomass consumption rate is shown in Fig. 4. It is found that with an increase in the air flow rate, biomass consumption rate increases. The increase in the air flow rate provides more oxygen to oxidize and higher amount of biomass would get combusted. The energy released will increase the rate of drying and pyrolysis. Biomass consumption rate increases not only due to a higher combustion rate, but also due to the enhanced pyrolysis and drying rate.

7.5 Performance Evaluation of Biomass Gasifier

The performance of the biomass gasifier system is evaluated in terms of the producer gas composition, the calorific value of producer gas, gas generation rate, zone temperatures and cold gas efficiency.

7.5.1 Producer Gas Composition

The gas composition of producer gas sampled at five minutes interval during gasification experiments is found using gas chromatograph (NUCON 5765) with thermal conductivity detector. The averaged gas composition for each experimental run is plotted against the equivalence ratio of the same experimental run and shown in Fig. 5. Molar fraction of nitrogen and carbon dioxide decreases with an increase in Φ upto a value of $\Phi = 0.205$ and for higher values of Φ, molar fraction of nitrogen and carbon dioxide increases. The fraction of carbon monoxide and hydrogen shows an increasing and decreasing trend just opposite to that of nitrogen and carbon dioxide.

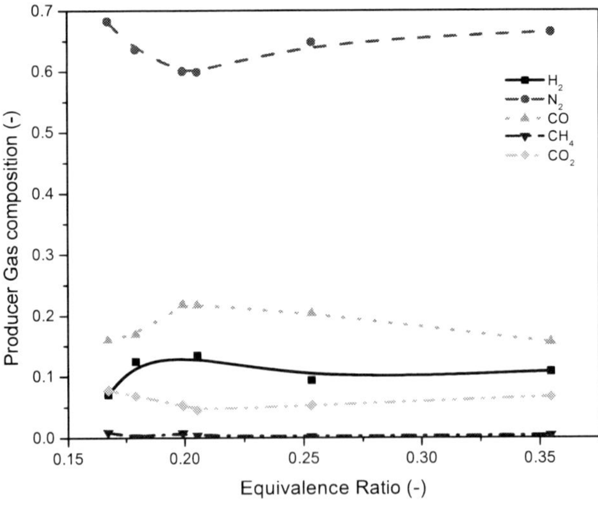

Figure 5. Effect of equivalence ratio on Producer gas composition.

A higher value of Φ represents a higher air flow rate for a specific biomass consumption rate which leads to more amount of CO_2 production in combustion zone and more amount of N_2 entry along with air flow. The conversion of CO_2 to CO depends upon the rate of reactions occurring in the reduction zone and length of the reduction zone. With an increase in Φ from 0.16 to 0.205, increased CO_2 amount in combustion zone is converted into carbon monoxide and hydrogen, and thereby the fraction of CO and H_2 increases with Φ till a value of Φ = 0.205 and fraction of CO_2 and N_2 decreases. The increase in CO_2 and decrease in CO & H_2 fractions for the equivalence ratio higher than 0.205 represents that CO_2 produced in combustion zone is in excess to that of the conversion capacity of reduction bed. The increase in N_2 fraction for the Φ value variation from 0.205 to 0.36 is due to a more amount of N_2 entry along with the air flow.

7.5.2 Zone Temperatures

Fig. 6 shows the variation of the average temperature of pyrolysis and oxidation zone of the downdraft biomass gasifier with the equivalence ratio. The temperature of oxidation zone varies from 900 °C to 1050 °C and that of pyrolysis zone between 260 °C to 550 °C. It clearly indicates that both temperature profiles pass through a maximum at an equivalence ratio of 0.205.

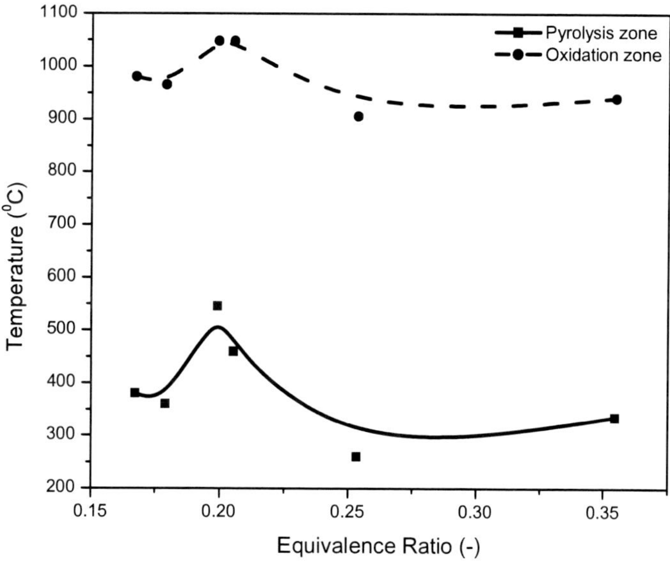

Figure 6. Effect of equivalence ratio on Zone temperatures.

The oxidation zone temperature depends upon the heat released due to the biomass combustion and air flow rate. As air flow rate increases, it provides more oxygen to oxidize but also brings inert N_2, which acts as a heat carrier and reduces the temperature of the oxidation and pyrolysis zone. The maximum value of temperatures in the pyrolysis and oxidation zones represents the optimum amount of equivalence ratio. Fig. 5 also supports that at an optimum equivalence ratio (Φ_{opt} = 0.205) the fractions of carbon monoxide and hydrogen are the maximum and the fraction of carbon dioxide is the minimum.

7.5.3 Calorific Value of Gas

The variation of calorific value of the producer gas with equivalence ratio is presented in Fig. 7. Calorific value is calculated using the composition of the producer gas. Carbon monoxide and hydrogen are the main components of the producer gas and are responsible for higher calorific value. It is found that at an equivalence ratio of 0.17 the calorific value is the least at 4.5 MJ/Nm^3. With a small increase in the equivalence ratio at $\Phi = 0.205$, the calorific value reaches to a maximum of 6.34 MJ/Nm^3 and then follows the decreasing trend. For the equivalence ratio higher than 0.205, calorific value steadily decreases with a further increase in the equivalence ratio. The increasing and decreasing trend of calorific value variation is exactly same as that of carbon monoxide and hydrogen variation with equivalence ratio as shown in Fig. 5.

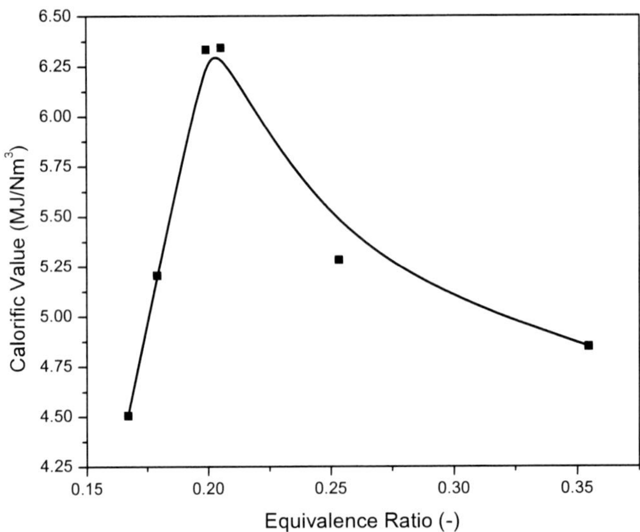

Figure 7. Effect of equivalence ratio on Calorific Value of producer gas.

7.5.4 Gas Production Rate

Fig. 8 shows the effect of equivalence ratio on the producer gas production rate per unit weight of biomass (N m^3/kg). It clearly shows that with an increase in the equivalence ratio, producer gas production rate continuously increases. Higher equivalence ratio signifies higher air flow rate for a specific biomass consumption rate. Hence although after a certain value of the equivalence ratio, calorific value of the producer gas deteriorates due to higher amounts of carbon dioxide as depicted in Fig. 5 and 7, the production rate of producer gas continue to increase.

7.5.5 Cold Gas Efficiency

Cold gas efficiency is defined as the ratio of energy of the producer gas per kg of biomass to the higher heating value (HHV) of the biomass material (Eq. (3)).

$$\text{Cold gas efficiency} = \frac{\text{(Calorific value)(Gas Production per weight of biomass)}}{\text{Higher Heating Value of the biomass}} \qquad (3)$$

The variation of cold gas efficiency with equivalence ratio is given in Fig. 8. As given in Eq. (3) cold gas efficiency depends upon the calorific value and the amount of producer gas released at constant HHV of biomass.

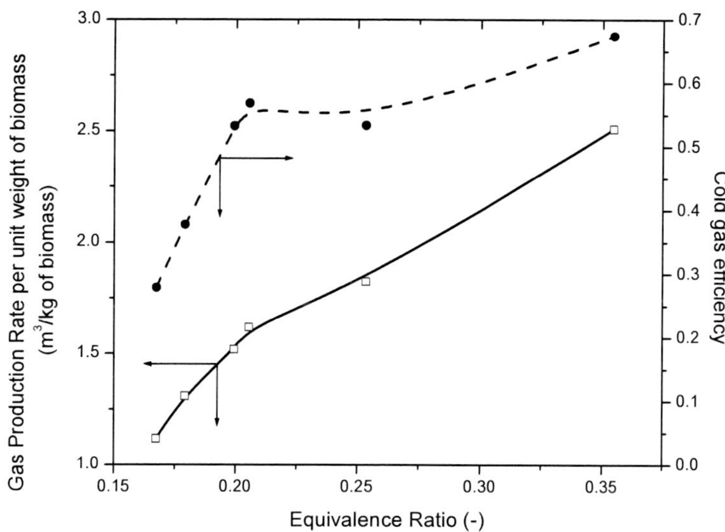

Figure 8. Effect of equivalence ratio on producer gas production rate per unit weight of biomass and on cold gas efficiency.

The amount of producer gas increases continuously and the calorific value passes through a maximum with an increase in the equivalence ratio. Cold gas efficiency is at a lowest value of 0.25 for an equivalence ratio value of 0.17. The value of cold gas efficiency becomes almost double with a small increase in the equivalence ratio at $\Phi = 0.205$. The effect Φ of on cold gas efficiency is comparatively lower for higher values of equivalence ratio. Cold gas efficiency increases from 0.5 to 0.6 for a change in the equivalence ratio from 0.2 to 0.35.

REFERENCES

Anthony, D.B., Howard, J.B., 1976. *Coal Devolatilization and Hydrogasification.* American Institute of Chemical Engineers Journal, 22, 625-656.

Babu, B.V., 2008a. *Biomass Pyrolysis: A State-of-the-art Review.* Biofuels Bioproducts and Biorefining, 2, 393-414.

Babu, B.V., 2008b. *Biohydrogen-A Clean and Green Energy: Economic Perspective & Optimization Strategies. Proceedings of Indo-French Workshop on 'Biohydrogen: from Basic Concepts to Technology' under the aegis of Indo French Center for Promotion of Advanced Research,* organized by The Energy and Resources Institute (TERI)-New Delhi, Jaypee Residency Manor, Mussoorie, November 6-8, 2008.

Babu, B.V., Chaurasia A.S., 2003a. *Modeling, Simulation, and Estimation of Optimum Parameters in Pyrolysis of Biomass.* Energy Conversion and Management, 44, 2135-2158.

Babu, B.V., Chaurasia A.S., 2003b. *Modeling for Pyrolysis of Solid Particle: Kinetics and Heat Transfer Effects.* Energy Conversion and Management, 44, 2251-2275.

Babu, B.V., Chaurasia, A.S., 2004a. *Parametric Study of Thermal and Thermodynamic Properties on Pyrolysis of Biomass in Thermally Thick Regime.* Energy Conversion and Management 45, 53-72.

Babu, B.V., Chaurasia, A.S., 2004b. *Dominant Design Variables in Pyrolysis of Biomass Particles of Different Geometries in Thermally Thick Regime.* Chemical Engineering Science 59, 611-622.

Babu, B.V., Chaurasia, A.S., 2004c. *Pyrolysis of Biomass: Improved Models for Simultaneous Kinetics & Transport of Heat, Mass, and Momentum,* Energy Conversion and Management 45, 1297-1327.

Babu, B.V., Chaurasia, A.S., 2004d. *Heat Transfer and Kinetics in the Pyrolysis of Shrinking Biomass Particle*, Chemical Engineering Science 59, 1999-2012 .

Barth, T., 1999. *Similarities and Differences in Hydrous Pyrolysis and Source Rocks.* Organic Geochemistry, 30, 1495-1507.

Bhave, A.G., 2001.*Technical notes in the Proceedings of Intensive workshop on - Testing Evaluation of Biomass Gasifier Systems and Related Laboratory Investigation.* Sardar Patel Renewable Energy Research Institute, VallabhVidhyanagar.

Bridgwater, A.V., Meier, D., Radlein, D. 1999. *An Overview of Fast Pyrolysis of Biomass.* Organic Geochemistry, 30, 1479-1493.

Bridgwater, A.V., 2002. *Bio-Energy Research Group, Aston University, Birmingham B47ET, UK,(2002).* http://www.icheme.org/.../conferences/gasi/Gasification% 20Conf%20Papers/ Session%202%20presentation-Bridgewater.

Caillé, A., 2007. *Survey of Energy Resources*. World Energy council, London, U.K.

Chaurasia, A.S., Babu, B.V., 2004. *Influence of Product Yields, Density, Heating Conditions, and Conversion on Pyrolysis of Biomass*. Journal of Arid Land Studies, 14S, 159-162.

Chowdhury, R., Bhattacharya, P., Chakravarty, M. 1994. *Modelling and simulation of a downdraft rice husk gasifier*. International Journal of Energy Research 18, 581-594.

Curtis, L.J., Miller, D.J. 1988. *Transport Model with Radiative Heat Transfer for Rapid Cellulose Pyrolysis*. Industrial and Engineering Chemistry Research, 27, 1775-1783.

Demirbas, A., Küçük, M.M., 1999. *Delignification of Ailanthus Altissima and Spruce Orientalis with Glycerol or Alkaline Glycerol at Atmospheric Pressure.* Cellulose Chemical Technology, 27, 679-686.

Demirbas, A., 2000. *Biomass resources for energy and chemical industry.* Energy Education and Science Technology 5, 21-45.

Demirbas, A., 2001. *Bioresource facilities and biomass conversion processing for fuels and chemicals*. Energy conversion and Management 42, 1357-1378.

Demirbas, A., Arin, G., 2002. *An Overview of Biomass Pyrolysis,* Energy Sources, 24, 471-482.

Di Blasi, C., Signorelli, G., Portoricco, G., 1999. *Countercurrent Fixed-Bed Gasification of Biomass at Laboratory Scale.* Industrial Engineering Chemistry and Research 38, 2571-2581.

Dogru, M., Howrath, C.R., Akay, G., Keskinler, B., Malik, A.A., 2002. *Gasification of hazelnut shells in a downdraft gasifier*. Energy 27, 415-427.

Fridleifsson, I.B., 2003. *Status of Geothermal Energy Amongst the World's Energy Sources.* Geothermics, 32, 379-388.

Jayah, T.H., Aye, Lu, Fuller, R. J., Stewart, D.F., 2003. *Computer simulation of a downdraft wood gasifier for tea drying.* Biomass and Bio energy 25, 459-469.

Jorapur, R.M., Rajvanshi, A.K. 1995. *Development of a sugarcane leaf gasifier for electricity generation.* Biomass and Bioenergy 8, 91-98.

Jorapur, R.M., Rajvanshi, A.K., 1997. *Sugarcane leaf-bagasse gasifiers for industrial heating applications*. Biomass and Bioenergy 13, 141-146.

Parikh, J., Channiwala, S.A., Ghosal, G.K., 2005. *A correlation for calculating HHV from proximate analysis of solid fuels.* Fuel 84, 487-494.

Reed, T.B., Das, A., 1988. *Handbook of Biomass Downdraft Gasifier Engine Systems,* The Biomass Energy Foundation Press, Colorado.

Sharma, A.K., 2007. *Modeling fluid and heat transport in the reactive, porous bed of downdraft (biomass) gasifier.* International Journal of Heat and Fluid Flow, 28, 1518–1530.

Shafizadeh, F., 1982. *Introduction to Pyrolysis of Biomass*. Journal of Analytical and Applied Pyrolysis, 3, 283-305.

Sheng, G.X., 1989. *Biomass gasifiers: from waste to energy production.* Biomass 20, 3-12.

Sheth, P. N. and B.V.Babu, 2009. *Experimental studies on producer gas generation from wood waste in a downdraft biomass gasifier.* Bioresource Technology, 100, pp. 3127-3133.

Singer, C.J., 1958. *History of Technology*, IV, 252, Claderon Press, Oxford.

Stucley, C.R., Schuck, S.M., Sims, R.E.H., Larsen, P.L., Turvey, N.D., Marino, B.E., 2004. *Biomass Energy Production in Australia,* A report for the Joint Venture Agroforestry Program (in conjunction with the Australian Greenhouse Office).

Twidell, J., 1998. *Biomass energy.* Renewable Energy World. 3, 38-39.

Wander, P.R., Altafini, C.R., Barreto, R.M., 2004. *Assessment of a small saw dust gasification unit.* Biomass and Bioenergy, 27, 467-476.

Zainal, Z.A., Ali, R., Quadir, G., Seetharamu, K.N., 2002. *Experimental Investigations of a downdraft biomass gasifier.* Biomass and Bioenergy, 23, 283-289.

In: Handbook of Research on Chemoinformatics ...
Editor: A.K. Haghi
ISBN: 978-1-62100-998-6

Chapter 6

ADVANCES IN OPTIMIZATION AND SIMULATION OF LOW DENSITY POLYETHYLENE (LDPE) TUBULAR REACTOR

***Ashish M. Gujarathi*[*] *and B.V. Babu*[**]**
Birla Institute of Technology and Science (BITS), PILANI-333 031
Rajasthan, India

1. LOW DENSITY POLYETHYLENE TUBULAR REACTOR

Low-density polyethylene (LDPE) is a thermoplastic made from ethylene with a density ranging from 0.915-0.935 g/cm^3. It was the first grade of polyethylene, produced in 1933 by Imperial Chemical Industries (ICI) using a high pressure process via free radical polymerization. Since the inception of production technology of LDPE, there have been a lot of changes in its process due to high demand of the quality finished product. It can be produced using a continuous process either in tubular or autoclave reactors at very high pressure (1300-3400 bars) and high temperatures (225-610 K). The high pressure polymerization using tubular reactors (Fig. 1) has gained popularity due to its better heat removal capacity, which helps in controlling the properties of the finished product. Some of the desirable properties of LDPE include low crystallinity, low density (0.915-0.935 g/cm3), and resistance to solvents, chemicals and oxidating agents, apart from the acceptable rheological behaviour (Kalyon et al., 1994). LDPE is widely used for manufacturing various high demand products such as containers, dispensing bottles, wash bottles, tubing, plastic bags for computer components, and various molded laboratory equipment. Due to these varied uses, there has been an increase in demand of LDPE. Agarwal and Han (1975) reported the effect of various operating parameters such as the axial mixing parameter (the Peclet number), change of feed conditions, the chain transfer to the dead polymers, etc. on the performance of reactor. Axial mixing analysis of tubular reactor was carried out by

*Lecturer, Chemical Engineering Group, Email: ashishg@bits-pilani.ac.in
**Dean-Educational Hardware Division & Professor of Chemical Engineering Email: bvbabu@bits-pilani.ac.in; Homepage: http://discovery.bits-pilani.ac.in/~bvbabu

incorporating the Peclet number in the model. However, Chen et al. (1976), in their study, observed the condition of excessive turbulence and concluded that the axial mixing must not be included due to a high value of Peclet number. Shirodkar and Taien (1986) presented a mathematical model for tubular reactor and compared the performance of the model with the actual plant data. Their predicted conversion and temperature profiles matched well with actual plant data within 10% accuracy. Brandoline et al. (1988) presented a mathematical model for high pressure tubular reactor for ethylene polymerization. The rate law parameters were evaluated using a set of experimental data available with them. They also reported the order of reaction to be 1.1 for the reaction of oxygen initiation. Dhib and Al-Nidawy (2002) carried out the modeling of free radical polymerization of ethylene using di-functional initiators in an autoclave reactor. The mechanism of di-functional initiation was investigated and the proposed model was tested with the conversion data of ethylene collected from the literature for one mono-functional initiator and two di-functional initiators. Goto et al. (1981) investigated the reaction mechanism of LDPE tubular reaction. They discussed the important experimental and modeling based aspects of polymerization describing the correlations of molecular structure, overall rate of polymerization, termination and propagation, chain transfer to solvent, backbiting, chain transfer to polymer, formation of unsaturated structure, etc. The values of rate law parameters corresponding to the above mentioned mechanism and the calculation method for molecular weight were reported.

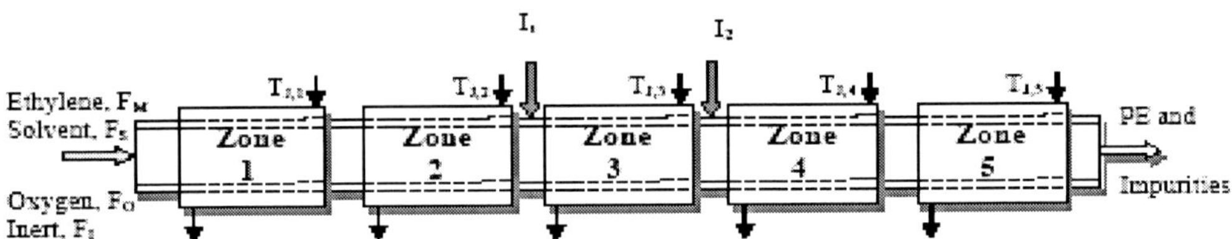

Figure 1. Schematic diagram of 5 zone tubular LDPE reactor.

A simulation based study of the single initiator feed tubular reactor showing the effect of several parameters such as change in initiator concentration, initial temperature, jacket temperature, reactor diameter, wall heat transfer coefficient, etc. on reactor performance was reported by Gupta et al. (1985). They showed the importance of such simulation based study and concluded by emphasizing the need for a detailed stability analysis considering that the inherent reactor instability can pose a considerable problem. Very few researchers (Lacunza et al., 1998) discussed the heat transfer aspects involved in the high pressure tubular reactor. They showed that the assumption of constant overall heat transfer coefficient (U) lead to a certain degree of error in conversion and the polymer property when compared with the experimental values. Lee et al. (2000) investigated the effects of change in initiator flow rate, and monomer flow rate, initiator concentration using a mathematical model for industrial high pressure autoclave polyethylene reactor including the decomposition phenomena. Brandoline et al. (1991) presented an optimization study and discussed the technical feasibility of each set of objectives with respect to the weight based multi-objective optimization scenario. Issues related to the effect of multiple feeds on temperature profile, and the trade-off between the conversion and the polydispersity index were discussed. The optimal control of the industrial reactor with and without initiator injections was carried out to maximize the final monomer conversion using the reactor jacket temperature as the control variable (2004).

Kalyon et al. (1994) presented a mathematical model and the rheological characterization of LDPE. Data from a commercial reactor having a reactor length of 720 m was used. They presented the results of their model predictions and the experimentally determined LDPE properties for long- and the short-chain branching for three different polyethylene resins. Rheological behavior of these different resins was studied with respect to apparent viscosity, shear stress, modulus of elasticity, etc. Recently, Kim and Iedema (2008) carried out the modeling of branching density and branching distribution in LDPE polymerization. They concluded that the concentration of long chain branching (LCB) is close to those of first branching moments in both the CSTR and the tubular reactor systems. Large variation in the rate law parameters is observed from the literature survey and the data observed in those studies is reported in Table 1.1a and Table 1.1b. The large variation in the values of rate law parameters is attributed to the complexity of the process and also to the variation in the geometry (length, number of initiator injections and its positions) of the reactor considered by the individual workers. Agarwal et al. (2006) used NSGA-IIaJG to solve LDPE tubular reactor under multi-objective optimization scenario. In this chapter an improved evolutionary optimization algorithm [Differential Evolution(DE)] and an improved numerical solver [numerical differentiation formulas (NDFs)] are used to estimate important reaction rate parameters and the operating variables which are responsible for analyzing the performance of LDPE A detailed parametric study for improved performance of LDPE tubular reactor is reported.

2. Simulation and Optimization

The reaction scheme (Eqs. A1 - A10), model equations (Eqs. A11 - A18), and the kinetic rate constants and equations related to LDPE tubular reactor (Eqs. A19 - A29c) considered in this study are given in Appendix. The design operating conditions are given in Table A1. Open literature contains several studies which incorporated several aspects related to modeling of continuous LDPE reactor.

But due to the complexity of reaction mechanism, large number of kinetic parameters and a wide range of experimental conditions over which the kinetic parameters are to be determined, the consistent set of rate constants has not been established. Table 1a shows the rate constant parameters in some of the studies reported in the literature. Table 1b shows the rate constant parameters used in the remaining studies (continued from Table 1a) and in the present study. Kiparissides et al. (1993) reported that under normal operating (experimental) conditions the values of propagation and termination (by combination) rate constants cannot be obtained. The reported values of rate constants by various authors (in Table 1a and 1b) also show a large deviation. Differential Evolution (DE), an evolutionary, population based search algorithm is found to be successful in handling many complex and non-linear engineering problems both in the field of single and multi-objective optimization (Price and Storn, 1997; Lee et al., 1999; Babu and Sashtry, 1999; Babu and Angira, 2005; Stumberger et al., 2000; Chakraborty et al., 2003; Babu, 2004; Babu et al., 2005; Babu and Munawar, 2007; Babu and Gujarathi, 2007; Gujarathi and Babu, 2009a-2009d). To bring down the deviation in the rate constant parameters and operating parameters, the differential evolution algorithm is used in the present study to obtain the optimum values of the parameters. The industrial values of temperature are read from the plot (Asteasuain et al., 2001), using a computer oriented 'scanit' software which can read the data with high accuracy.

The data is read at 28 discrete points from the plot and are shown in Table 2a along with the model predicted data at a specified length. Apart from the temperature at 28 discrete points, the exit values of monomer conversion, the side product concentration and the number-average molecular weight are used for minimizing function, *I*. Brandoline et al. (1996) presented a comprehensive model that proved to be a well known model which included several mechanisms of termination reactions.

Table A1. Operating conditions used for industrial tubular LDPE reactor in the present study for reference run (Brandoline et al., 1996; Agarwal et al., 2006)

Parameter	Value
Feed temperature, T_{in}	349.15 K
Feed pressure P_{in}	227.98 Mpa
Initiator 1 feed flow rate, F_{I1}	1.0 x 10^{-3} kg/s
Initiator 2 feed flow rate, F_{I2}	1.6 x10^{-4} kg/s
Feed flow rate of oxygen, F_{O2}	6.8 x 10^{-5} kg/s
Feed flow rate of solvent, F_S	7.4 x 10^{-2} kg/s
Feed flow rate of inert, F_{inrt}	0.22 kg/s
Feed flow rate of jacket fluid, $V_{jm\ (m=2,...5)}$ Feed flow rate of jacket fluid, $V_{jm\ (m=2,...5)}$	6.90e^{-3}, 6.34e^{-3},1.67e^{-3},1.25e^{-3} m^3/s (Case A)* 6.73e^{-3}, 3.99e^{-3},0.978e^{-3},2.78e^{-3} m^3/s (Case B)*
Wall heat transfer coefficient, h_W	1256 W/m^2C
Feed conditions for live and dead moment, λ_{np} μ_{np} (n=0,1; p=0,1,2)	0.0 $kmol/m^3$
Specific heat of reaction mixture, C_{pm}	2.42834, 2.42834, 3.1401, 3.1401, 4.01933 kJ/kg K
Mean jacket temperature, $T_{j,m}$ (m=1,...,5)	441.15, 498.15, 498.15, 441.15, 441.15 K
Feed velocity v	10.14 m/s
Total reactor length, L_t	1390 m
Inside diameter of reactor, D_t	0.05 m
Thickness of reactor wall, t	0.0254 m
Inner diameter of outer (jacket) wall D_{Ji}	0.2032 m
Axial lengths of reactor zones , L_{zn} (n=1,2,..5)	60, 100,180, 510, 540 m

*Results obtained in present study

Agarwal et al. (2006) made use of a model proposed by Brandoline et al. (1996). They carried out the tuning of important rate parameters involved in the model. When we tried to use the model predicted parameters reported by Agarwal et al. (2006) in the present study simulation runs using MATLAB, we observed an average deviation of 0% in X_m, 14.36 % in [*SCB*] and 12.58 % in M_n respectively at the exit of reactor. However, the trends obtained for each species in our study remain the same, as reported in the literature (Brandoline et al., 1996; Agarwal et al., 2006). The important equations (such as friction factor) related to the heat transfer operations were not reported in the literature (Agarwal et al., 2006). Therefore we considered two different cases of heat transfer in this study. In Case A, Eq. A29a, which represents the explicit form of friction factor (Brandoline et al., 1991) is used. In case B, an implicit form of friction factor (Eq. A29b and A29c), as reported by Kiparissides et al. (1993) is considered.

Table 1a. Rate constant parameters and bounds used in some of the studies reported in the literature

Reaction	Parameter	Bounds of variables considered in this study	Values reported by Goto et al. [7]*		Values reported by Chen et al. [3]**		Values reported by Agarwal and Han [2]*		Values reported by Shirodkar and Tsien [4]*	
			A_0 (L/mol.s)	E (cal/mol)	A_0 (L/mol,s)	E (cal/mol)	A_0 (L/mol. s)	E (cal/mol)	A_0 (L/mol.s)	E (cal/mol)
Initiation	E_o	$125604<E_o<138164$	2.39×10^{19}	37300 + 0.06P	1.6×10^{16}	38400	$0.75k_{d0}$	E_d+0.17P	--	--
Peroxide Initiation 1	E_{d1}	$117230<E_{d1}<136071$	--	--	--	--	--	--	--	--
Peroxide Initiation 2	E_{d2}	$117230<E_{d2}<133977$	--	--	--	--	--	--	--	--
Propagation	E_p	$14653<E_p<18003$	5.63×10^{11}	10520-0.477P	2.95×10^{7}	7091	1.25×10^{8}	7800+0.5P	5.8×10^{7}	7,769-0.259P
Chain transfer to solvent/ modifier	E_{trs}	$14653<E_{trs}<20934$	1.23×10^{19}	12820-0.4722P	--	--	$k^0_{trs}\exp(12000/RT^0)$	12000	--	--
Backbitting (Intramolecular chain transfer)	E_{bb}	$56521<E_{bb}<66988$	5.63×10^{12}	13030-0.569P	--	--	--	--	--	--
β-Scisson of seco. radical	E_{b1}	$71175<E_{b1}<87922$	8.51×10^{10}	14530-0.477P	2.72×10^{11}	20000	--	--	--	--
β-Scisson of tert. radical	E_b	$62802<E_b<87922$	5.82×10^{11}	15760 –0.547P	--	--	--	--	--	--
Termination by combination	E_{bc}	--	3.00×10^{11}	3000 +0.3418P	1.6×10^{9}	2400	2.2×10^{10}	1000+0.244P	2.8×10^{8}	298+0.012P
By thermal degradation	E_{tdt}	--	--	--	--	--	--	--	--	--
By chain transf. to polymer	E_{trp}	--	1.75×10^{12}	14080+0.1065P	9.0×10^{5}	9000	$k^0_{trp}\exp(12000/RT^0)$	12000	7.5×10^{6}	8942+0.019P

*E in cal/mol, P in atm, $k=A_0 \exp(-E/RT)$, **E at 2000 atm;

Table 1b. Rate constant parameters and bounds used in the remaining studies (continued from Table 1a)

Reaction	Parameter	Values reported by Brandoline et al. [5]*		Values reported by Lee et al. [10]*		Values reported by Kim and Iedema [13]*		Values reported by Brandoline et al [35]#		Values reported by Agrawal et al. [14]#
		A_0 (L/mol.s)	E (cal/mol)	A_0	E (cal/mol)	A_0	E (cal/mol)	A_0	E (kJ/kmol)	E (kJ/kmol)
Initiation	E_o	$3.0x10^{10}$	27,941	--	--	--	--	$^{\$}1.6x10^{16}$	135945	132168
Peroxide Initiation 1	E_{d1}	--	--	$^{@}1.309x10^{19}$	33,872	$^{\&}1.35x10^{13}$	14130+0.033P	$^{\theta}1.0x10^{14}$	94621-133140	119929
Peroxide Initiation 2	E_{d2}	--	--	$^{@}1.396x10^{13}$	30,103	$^{\&}2.89x10^{14}$	16627+0.1217P	$^{\theta}1.0x10^{12}$	94621-132721	123117
Propagation	E_p	$1.0x10^{6}$	5,245	$^{\Delta}4.16x10^{6}$	6477-0.56P	$^{\theta}1.88x10^{7}$	4125-0.324P	$^{\theta}1.6x10^{16}$	17626	17431
Chain transfer to solvent/ modifier	E_{trs}	$1.7x10^{6}$	9,443	$^{\Delta}1.309x10^{19}$	33,872	$^{\theta}1.99x10^{7}$	5499-0.3253P	$^{\theta}4.0x10^{5}$	17253	18406
Backbitting (Intramolecular chain transfer)	E_{bb}	--	--	--	--	--	--	$^{\theta}1.6x10^{16}$	61964	60537
β-Scisson of seco. Radical	E_{b1}	--	--	-	--	--	--	$^{\theta}1.6x10^{16}$	79967	84747
β-Scisson of tert. radical	E_b	--	--	--	--	--	--	$^{\theta}1.6x10^{16}$	79967	70205
Termination by combination	E_{bc}	$3.0x10^{8}$	3,950	$^{\Delta}3.0x10^{8}$	3.950	$^{\theta}8.11x10^{8}$	553.26+0.19P	$^{\theta}1.6x10^{16}$	15282	15282
By thermal degradation	E_{tdt}	$7.3x10^{6}$	11,315	$^{\Delta}3.0x10^{8}$	3.950	--	--	$^{\theta}1.6x10^{16}$	79968	79968
By chain transf. to polymer	E_{trp}	$4.4x10^{6}$	9,500	$^{\Delta}3.0x10^{4}$	9375-0.48P	$^{\theta}2.15x10^{5}$	5921-0.04059P	$^{\theta}1.6x10^{16}$	36844	36844

*E in cal/mol, P in atm, $k=A_0 \exp(-E/RT)$, **E at 2000 atm; $^{\theta}A_0$ in m^3/kmol.s; $^{\&}A_0$ in m^3/s; $^{@}A_0$ in L/mol.s; $^{\Delta}A_0$ in s^{-1}

#$k=A_0 \exp(-(E+10^3 P\Delta V)/RT)$, where, P in Mpa, E in kJ/kmol, ΔV in m^3/mol, $A_0=m^3kmol^{-1}s^{-1}$, $ $m^{3.3}kmol^{-1}s^{-1}$

[a] Asteasuain et al. [34]

Table 2a. Comparison of model predictions with the industrial data on temperature profile along the axial length of the reactor

Axial Length (m)	Temperature, T (K)		Axial Length (m)	Temperature, T (K)	
	Industrial data (Asteasuain et al., 2001)	Model predicted [present study]		Industrial data (Asteasuain et al., 2001)	Model predicted [present study]
0	350.0	350.00	603	529.2	528.27
80	390.1	386.89	640	520.1	522.02
113	407.8	402.99	677	514.4	516.38
139	421.3	414.09	720	507	510.32
173	439.9	431.14	762	499.2	504.89
180	450.9	437.56	810	492.6	499.20
209	516.4	531.68	857	503.7	497.42
273	588.8	610.97	904	543.3	533.35
307	592.7	600.36	940	562.3	555.54
350	584.9	586.58	980	568.8	570.19
388	575.4	576.03	1034	563.9	566.00
430	563.8	565.27	1110	560.4	557.00
463	553.5	555.63	1176	556	550.00
510	546.1	545.51	1254	547.5	541.00
549	539.1	537.81	1329	542.7	533.86

The comparison of industrial data and the model predictions (for both case A and case B) is given in Table 2b. In terms of monomer conversion, model predictions are exactly matching with industrial data in both the cases (i.e., case A & case B). Case B model predicted the exact value of number-average molecular weight, whereas Case A model predicted the value of M_n with a relative error of 0.48%. Both Table 2a and 2b show that with an efficient numerical solution procedure (NDFs), the model predicted values of present study matches well with the industrial data. The difference in parameter values obtained in the present study and in the previous study (Agarwal et al., 2006) may be attributed to the two different solution procedures used in these studies. Considering the much accurate estimation of model output, Case B is used for simulation in this study.

Table 2b. Comparison of model predictions of present study with the reported industrial data (Agarwal et al., 2006) of various properties at the exit of the reactor

Specific property at the exit of reactor	Industrial	Model Predicted values		
		Using Gears Routine and NSGA (Agarwal et al., 2006)	Using Numerical differentiation formulas (NDFs) and DE (Present Study)	
			Case A	Case B
M_n (kg/kmol)	21900*	21901	21793	21900
X_M	0.3*	0.2971	0.3	0.3
SCB (per 10^3 C atom)	30$^{\Delta}$	30.13	29.63	29.91
V_i (per 10^3 C atom)	0.1§	0.1	0.099	0.099
V_{id} (per 10^3 C atom)	0.7$^{\Delta}$	0.7	0.684	0.695

*Data taken from Asteasuain et al. (2001); §Data taken from Gupta et al. (1985); ΔData taken from Goto et al. (1981)

3. Simulation and Parametric Estimation using Differential Evolution

ODE15s subroutine of MATLAB (7.0) library is used in the present study. Agarwal et al. (2006) used the Gears Routine (D02EJF of NAG library) in their simulation runs. ODE15s (in MATLAB 7.0 library) is a variable order solver based on the numerical differentiation formulas. ODE15s has an option to use backward differentiation formula (BDFs, also known as Gears method) that is usually less efficient. NDF can achieve the same accuracy as BDF with step size about 26% bigger (Shampine and Reichelt, 1997). This clearly highlights the inefficiency of BDF (Gears routine) in comparison with NDF. The slight deviation in the exit concentration of various species [obtained using the parameters values reported in literature (Agarwal et al., 2006)] encouraged us to fine tune the model in order to find the rate law parameters associated with the eight reactions of initiation, propagation and termination.

Differential Evolution (DE) algorithm is used to minimize the sum of square of the normalized error, I, between the model-predicted values and the industrial values (Eq. 5.1).

$$Minimize \quad I(u) = \sum_{i,j}\left(1 - \frac{N_i^{ind}(z_j)}{N_i^{model}(z_j)_{exit}}\right)^2 \tag{5.1}$$

where, N_i is the value of ith property, and the superscripts m and "ind" represent the values predicted by the model and the industrial values, respectively. Properties used in the estimation of function I in this study are temperature of reactor, number-average molecular weight, monomer conversion, and methyl, vinyl and vinylidene end groups per 1000 carbon atoms in the chain. The decision variables involved in this study are given by Eq. 5.2.

$$I(u) \cong f(E_o, E_{d1}, E_{d2}, E_p, E_{trs}, E_{bb}, E_{b1}, E_b, V_{j2}, V_{j3}, V_{j4}, a_v) \tag{5.2}$$

The internal heat transfer coefficient also affects the reaction kinetics and therefore the jacket fluid flow rates and parameter involved in viscosity estimation were also incorporated in the list of decision variables as shown in Eq (5.1). The final iteration of DE algorithm resulted in several near global solutions with an accuracy of the order of 10^{-4}. Out of those near global solutions, the set of optimal decision variables are randomly selected and they are reported (for both case A and case B) in Table 3 along with the optimum values reported by Agarwal et al. (2006), operating plant values reported by Brandoline et al. (1996) and the bounds of the parameters. Brandoline et al. (1996) provided the values of initiator concentration in the form of ranges due to certain proprietary reasons. Initial initiator concentration plays an important role in controlling the performance of reaction scheme (especially in the presence of varied set of input conditions). Table 3 shows that most of the optimum parameters obtained in this study are close to those obtained by Brandoline et al. (1996). As the model predictions are matching well with the industrial data (Table 2b), a detailed parametric study is carried out through extensive simulations by varying different affecting decision variables and design parameters. In the next sub-sections, the results of simulation of LDPE tubular reactor are discussed by considering the effect of various feed conditions and the operating design parameters.

Table 3. Bounds, reported values and final tuned values (present study) of the parameters

Parameter	Bounds	Value of the parameter			
		Reported (Brandoline et al., 1996)	Using Gears method and NSGA (Agarwal et al., 2006)	Using Numerical differentiation formulas (NDFs) and DE (Present study)	
				Case A	Case B
E_o	$125604 < E_o < 138164$	135945	132168	133232	134892
E_{d1}	$117230 < E_{d1} < 136071$	94621-133140	119929	123702	130189
E_{d2}	$117230 < E_{d2} < 133977$	94621-132721	123117	127803	122373
E_p	$14653 < E_p < 18003$	17626	17431	18002	17809
E_{trs}	$14653 < E_{trs} < 20934$	17253	18406	20927	20930
E_{bb}	$56521 < E_{bb} < 66988$	61964	60537	60832	60591
E_{b1}	$71175 < E_{b1} < 87922$	79967	84747	79828	79809
E_b	$62802 < E_b < 87922$	79967	70205	76149	75959
V_{j2}	$0.005 < V_{j2} < 0.007$	0.0012	0.00403	0.00690141	0.00673415
V_{j3}	$0.005 < V_{j3} < 0.007$	0.0012	0.00394	0.00634553	0.00399092
V_{j4}	$0.005 < V_{j4} < 0.007$	0.0012	0.00332	0.00167684	0.00097869
V_{j5}	$0.001 < V_{j5} < 0.005$	0.0012	0.00022	0.00125074	0.00278861
a_v	$0.009 < a_v < 0.0185$	0.017	0.018	0.0132271	0.01477835

Results and Discussion

Considering the improved kinetic and operating parameters obtained in this study, a detailed parametric study is carried out to check the effects of individual parameters on the performance of LDPE tubular reactor.

3.1 Effect of Feed and Jacket Temperature on the Reactor Performance

The effect of feed and jacket temperature is analyzed by plotting the axial variation of temperature, short chain branching (SCB), vinyl and vinylidene end group per 1000 CH_2 atoms along the length of the reactor as shown in Figs. 2-5.

The data is analysed by varying one variable, while fixing the values of other variables constant at reference values (Table D1 in appendix D). Table 4 shows the effect of change in feed and jacket temperature on various aspects [such as, T_{MAX}, L_a, M_w, M_n, PDI, X_m and the exit concentrations of side groups (i.e. SCB, vinyl and vinylidene groups per 1000 CH_2 atoms as given by Eqs. A23-A29)].

Table 4 shows that the maximum temperature attained in the reactor and the length at which it is attained play an important role in deciding the overall quality of the polymer. Polydispersity index (PDI), which is a measure of the distribution of polymer molecular mass in a given polymer sample, depends on these two factors. As per the kinetics, a lower value of jacket temperatures delays the propagation reaction. Fig. 2 shows that the maximum attained temperature (T_{MAX}) and its location depend on the temperature of jacket fluid. As the initiator concentration gets depleted the temperature of the reaction mass increases. Once the maximum temperature in the reactor is attained, the monomer conversion ceases and the properties such as, M_n, ρ, [*SCB*], [*vinyl*] and [*vinylidene*] concentration remain constant until the addition of another initiator at a length of 850 m along the length of the reactor (Figs. 3 - 5). The concentrations of side chains ([*SCB*], [*vinyl*] and [*vinylidene*]) depend on the maximum temperature attained in the reactor. Higher the value of peak temperature attained in the reactor, the higher is the concentration of undesired side chain species. Sr. No. 4 of Table 4 corresponds to the lowest value of jacket temperatures (383 K in zones 1, 4, 5; and 430 K in zones 2 and 3). The low heating rate resulted in lowest value of maximum temperature attained among the 5 different cases considered in Table 4. The low value of jacket fluid temperature also resulted in a greater reactor length (Fig. 2 and L_a in Table 4) to attain the maximum temperature. This operating condition however resulted in low values of unwanted side chain concentration with an enhanced value of polydispersity index.

Sooner the peak temperature achieved in the reactor length, the smaller is the concentration of undesired side chains. The lower bound of jacket temperature (i.e., Sr. No. 4 in Table 4) due to low heating rate causes shortest distance of peak temperature, resulting in the lowest value of side chain concentrations per 1000 carbon atoms (i.e., SCB = 26.26, vinyl end group = 0.083846 and vinylidene end group = 11.891). The reaction mixture is cooled in the fourth zone in order to obtain a maximum efficiency of the initiator. It is also interesting to note that maximum temperature attained in the reactor and the concentration of side chain products are relatively insensitive to the feed temperature over the ranges covered in this study. This result is consistent with those reported by Gupta et al. (1985). However, the changes in the values of jacket temperature show the effect on these parameters as shown in Table 4. PDI also depends on the jacket temperature and feed temperature. Smaller the value of jacket temperature and feed temperature (considered in this study), the greater is the value of PDI.

Undesired side chain profiles (Figs. 3 - 5) show that the concentration of SCB, vinyl and vinylidene groups are observed immediately after the initiator injection in zone 3. The concentration of these species remains almost constant until the second initiator is injected in the fifth zone. The change in concentration of unwanted side products in zone 5 is less abrupt as compared to that observed in zone 3. This could be due to the high concentration of already formed polymer in zone 3, 4 and at the beginning of zone 5.

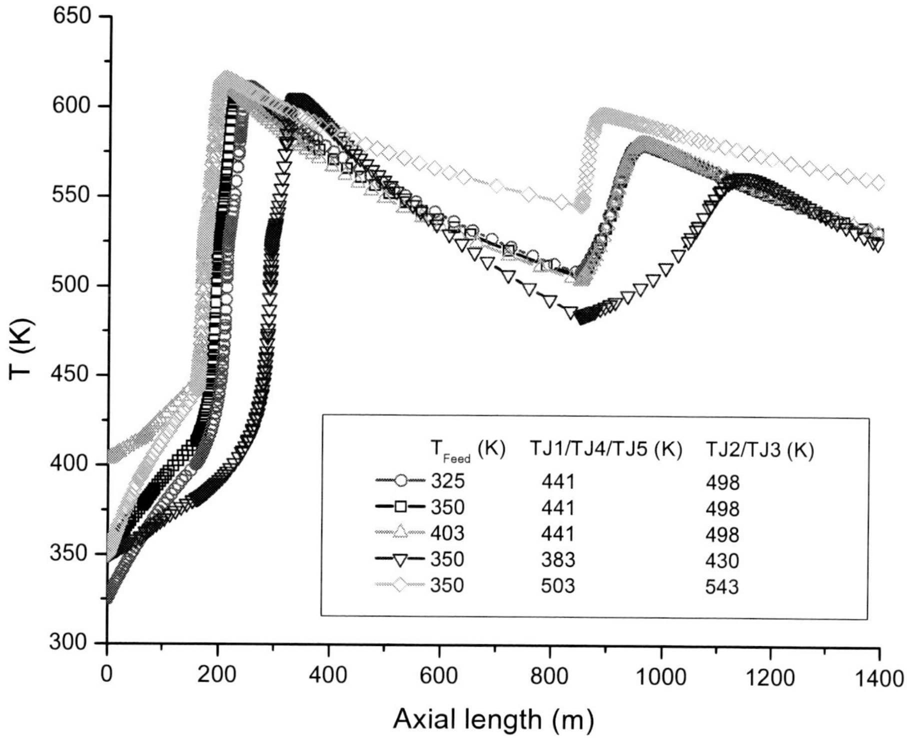

Figure 2. Effect of jacket temperature on the temperature profile along the reactor length.

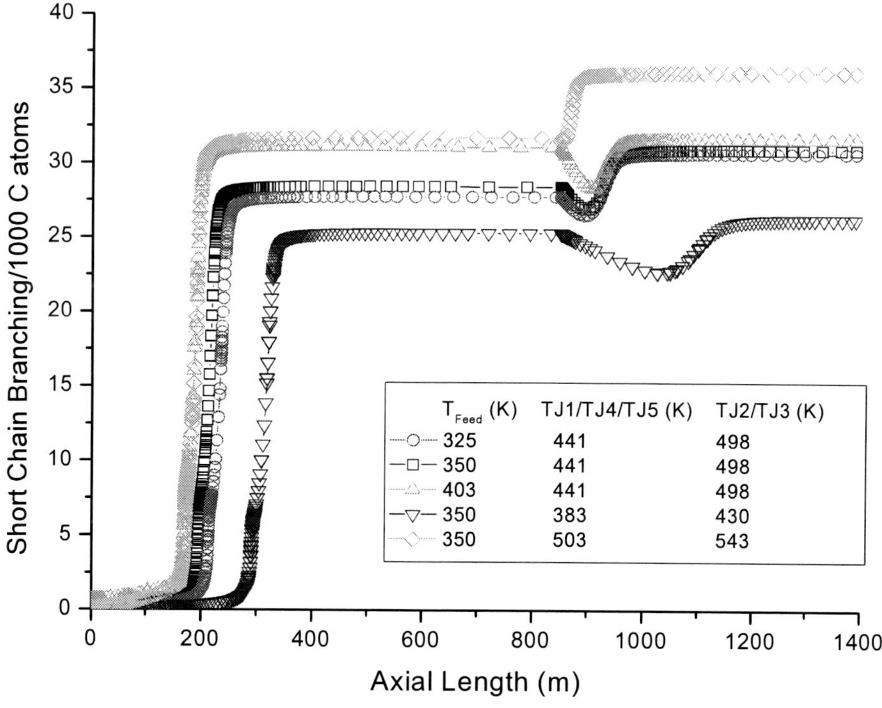

Figure 3. Effect of feed and jacket temperature on Short chain branching content along the reactor length.

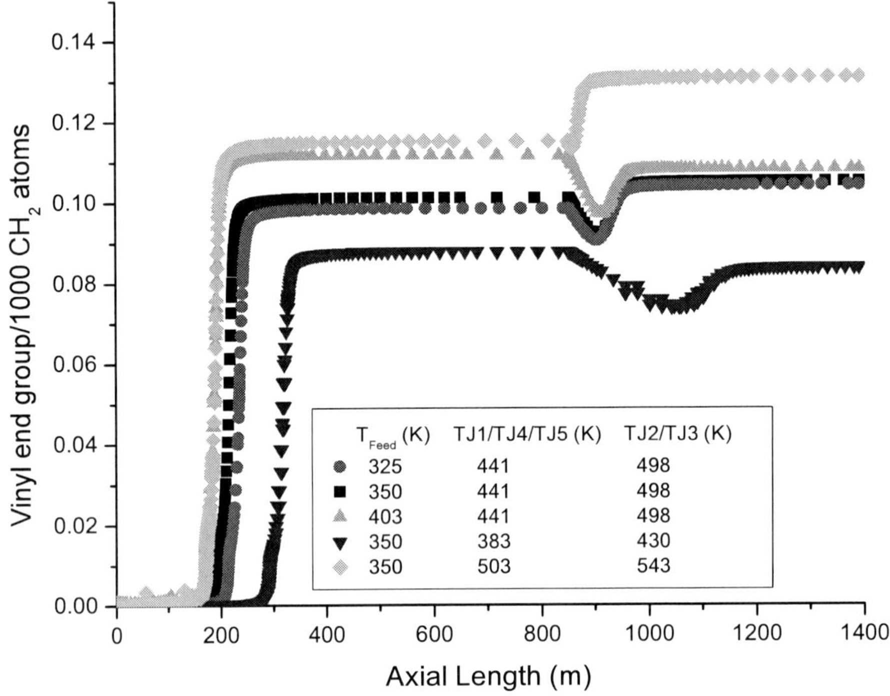

Figure 4. Effect of feed and jacket temperature on vinyl end group profile along the reactor length.

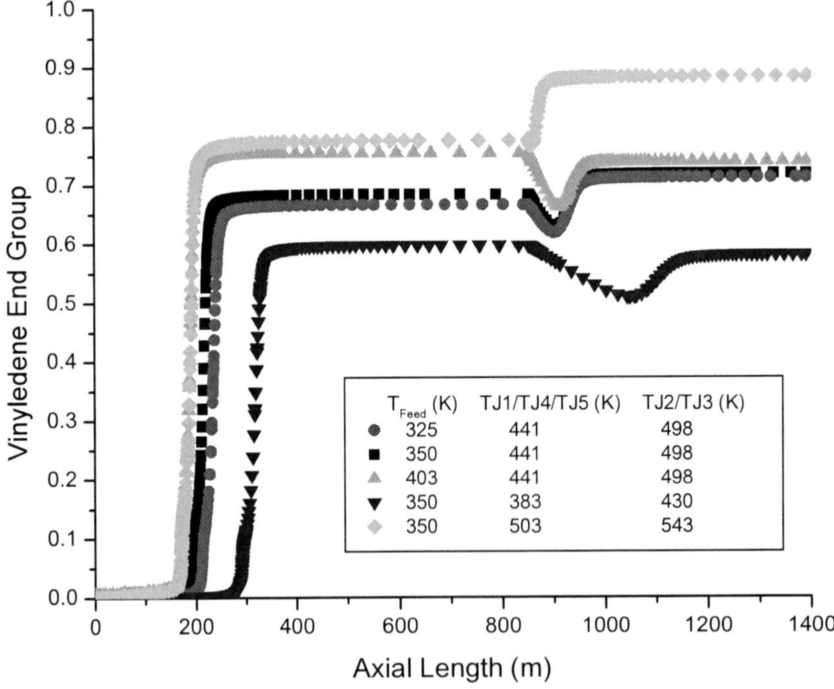

Figure 5. Effect of feed and jacket temperature on vinylidene end group profile along the reactor length.

Table 4. Various properties and exit concentration of different functional groups at various feed and reactor jacket temperatures

Sr. No.	T_{Feed} (K)	Jacket temperature		Max. Temp T_{MAX} (K)	Axial Position at Max Temp L_a (m)	M_w (kg/kmol)	M_n (kg/kmol)	PDI	X_m (%) (Exit)	Exit Side chain concentration/10^3 C atoms			Unconverted Ethylene (kmol/m^3)
		T_1 (K)	T_2 (K)							Methyl Group (SCB)	Vinyl Group	Vinylidene Group	
1	325	441	498	611.29	231.8	295175	21004	14.053	0.316	30.66874	0.10419	0.712635	13.61
2	350	441	498	611.29	251.9	242368	20827	11.636	0.310	30.91519	0.105093	0.718724	13.796
3	403	441	498	612.86	204.8	162743	20245	8.0384	0.2932	31.60865	0.108265	0.73925	14.616
4	350	383	430	605.18	334.7	554464	25388	21.8393	0.3922	26.26394	0.083846	0.580967	11.891
5	350	503	543	615.66	207.3	133493	17064	7.8230	0.2444	36.10889	0.130979	0.883567	15.7839

T_1 = Jacket temperature in zones 1, 4, and 5; T_2 = Jacket temperature in zones 2 and 3

3.2 Effect of Initiators Concentration on the Reactor Performance

Figs. 6 – 10 show the effect of change in the injected initiator concentration on the axial variation of various entities along the length of the reactor. Initiator I_1 and I_2 are injected at a length of 160 m and 850 m from the entrance respectively. The effect of initiator injections on temperature profile along the reactor length is shown in Fig. 6a.

For clarity, Fig. 6a is magnified in Fig. 6b and Fig. 6c. Both figures are magnified in such a way that the region of initiator injection (150-300m and 800-1250m) is clearly visible. Lower the concentration of initiator I_1, the greater is the length it acquires for reaching the maximum temperature (see Fig. 6a). The peak attained in the temperature marks the depletion of initiator concentration. Thereafter, not much happens in the reactor, except the cooling of reaction mixture. The concentrations of side chains ([SCB], [vinyl] and [vinylidene] end groups) also depend on the peak attained by temperature in the respective regions (Figs. 7 - 9). The peaks attained in these regions in turn depend on the extent of propagation reaction that occurs in the respective regions in the reactor. Figs. 6b-6c and Table 5 show that the length required attaining the maximum temperature in the reactor, in both the zones, increases with an increase in the concentration of initiators in the respective zones. The studies reported by Gupta et al. (1985), Brandoline et al. (1988), and Shirodkar and Taien, (1986) show a single peak in the profiles of temperature, unwanted side chain concentrations, and the monomer conversion. However, in this study and other studies reported in the literature (Asteasuain et al., 2001; Agarwal et al., 2006), multiple peaks of temperature, side chain concentrations and monomer conversion are observed. The number of peaks achieved in a reactor depends on the number of injections. The exit monomer conversion remains nearly independent of initiator concentration over the ranges considered in this study (Fig. 10). However, the number-average molecular weight decreases with an increase in the values of initiator 1 and initiator 2 over the ranges reported in Table 5. The value of Reynolds number gradually decreases along the post peak temperature region (Fig.11). The initiator injection marks the start of highly exothermic reaction, thus increasing the values of Reynolds number and temperature.

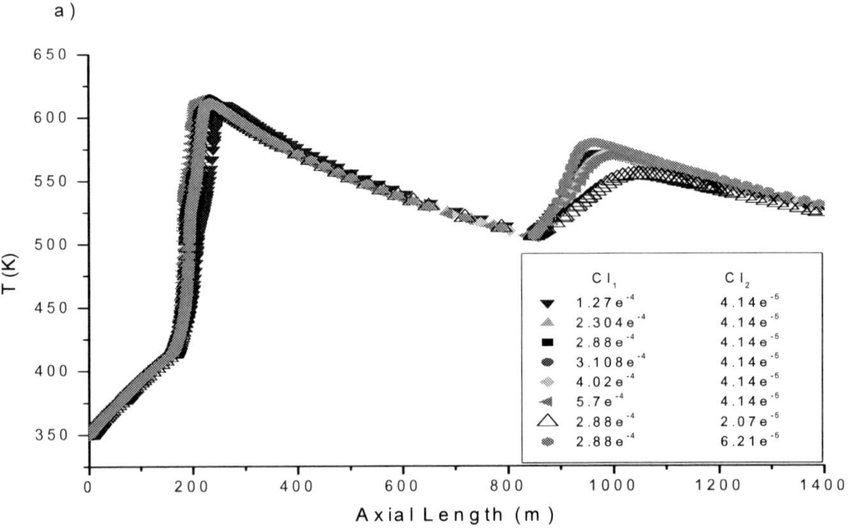

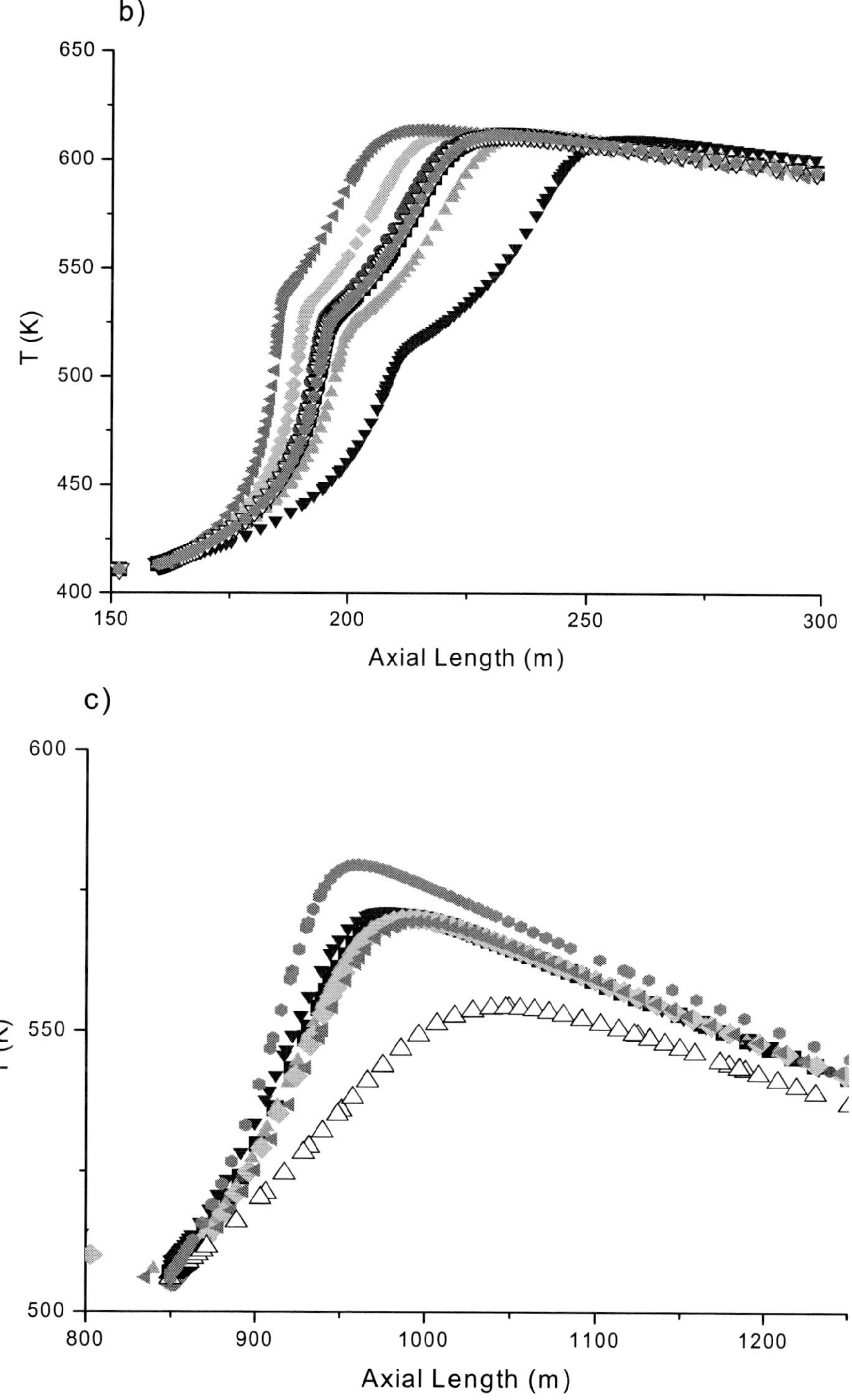

Figure 6. (a) Effect of initiator concentration on the reactor temperature profile; (b) Magnified part of Figure (a) in the range of 150-300 m reactor length (c) Magnified part of Figure (a) in the range of 800-1250 m reactor length.

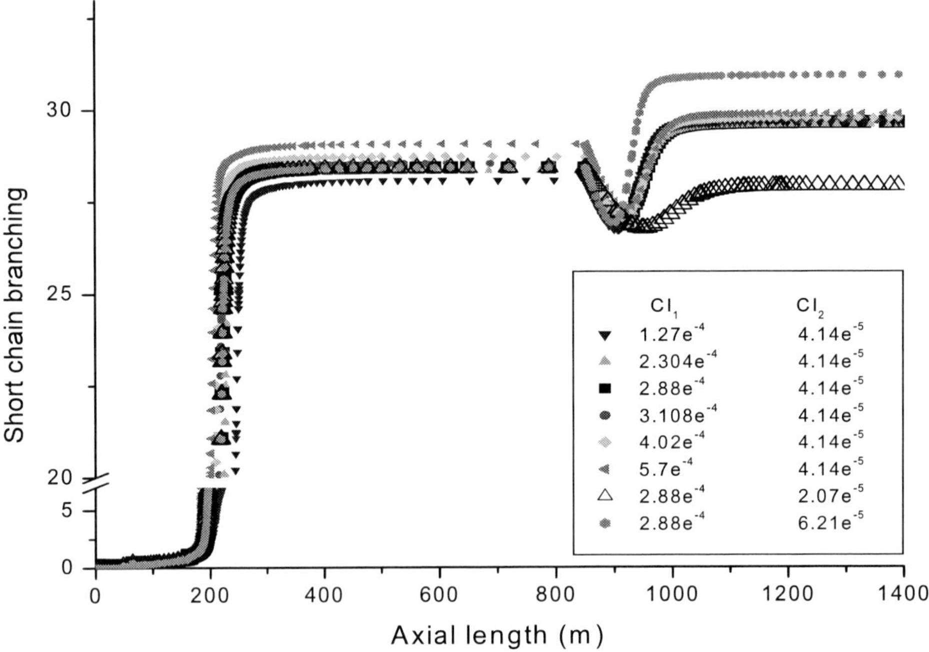

Figure 7. Effect of initiator concentration on Short chain branching content profile along the reactor length.

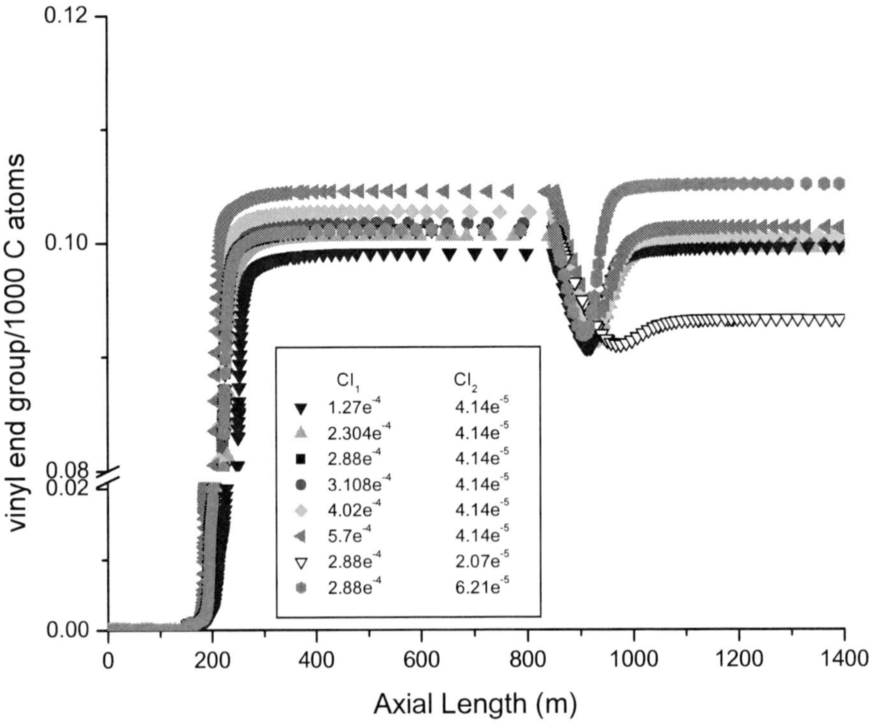

Figure 8. Effect of initiator concentration on vinyl end group per 1000 CH_2 atoms profile along the reactor length.

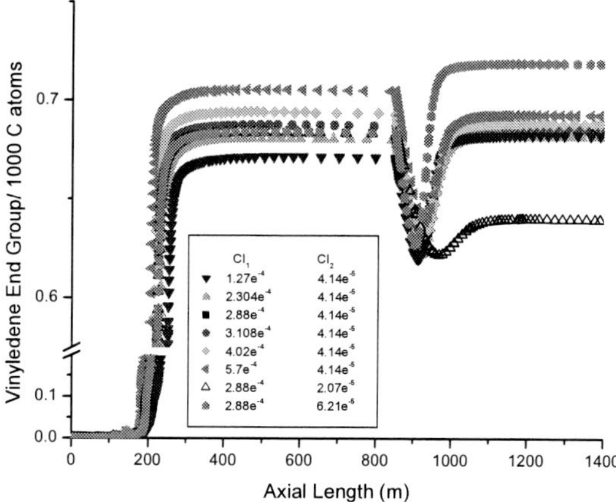

Figure 9. Effect of initiator concentration on vinylidene end group per 1000 CH_2 atoms profile along the reactor length.

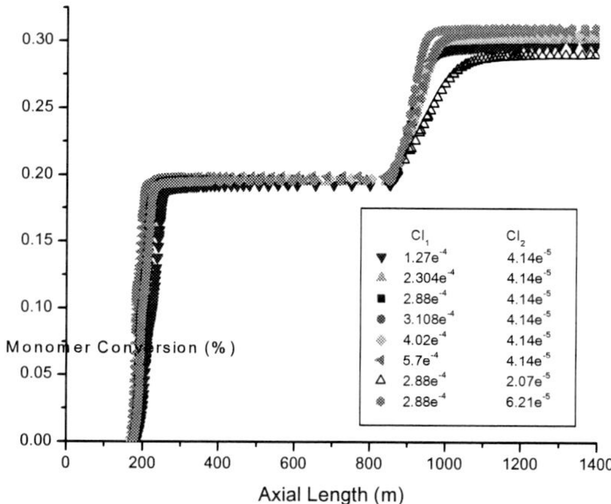

Figure 10. Effect of initiator concentration on Monomer conversion along the reactor length.

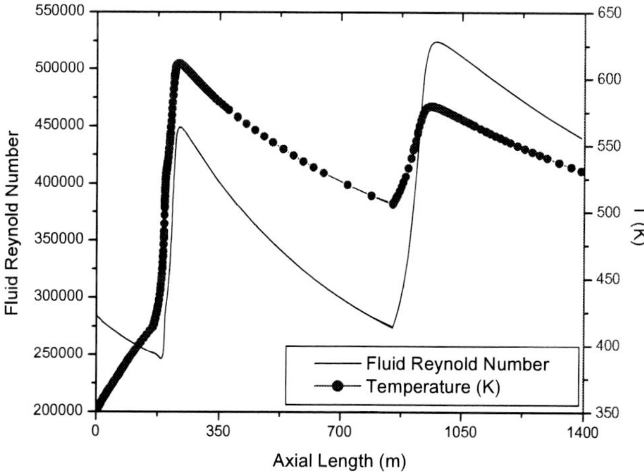

Figure 11. Fluid Reynolds number and temperature profile of tubular LDPE reactor along the reactor length.

Table 5. Various properties and exit concentrations of various species at different values of initiator concentration

Sr. No	CI_1 (kmole/m^3)	CI_2 (kmole/m^3)	Max Temp. T_{MAX} (K)	Axial Position at Max Temp., L_a (m)	M_w (kg/kmol)	M_n (kg/kmol)	PDI	Xm (%) (Exit)	Side chain concentration/10^3 C atoms			Unconverted Ethylene (kmol/m^3)
									Methyl Group (SCB)	Vinyl Group	Vinylidene Group	
1	$1.270e^{-4}$	$4.14e^{-5}$	609.99	258.73	280599	22286	12.59	0.2964	29.64	0.0995	0.6825	14.05
2	$2.304e^{-4}$	$4.14e^{-4}$	611.16	239.25	256053	22020	11.62	0.302	29.57	0.099	0.6817	13.96
3	$2.880e^{-4}$	$4.14e^{-5}$	611.63	231.82	243594	21792	11.17	0.302	29.63	0.09985	0.6807	13.96
4	$3.168e^{-4}$	$4.14e^{-5}$	611.93	229.63	238116	21668	10.98	0.303	29.68	0.100	0.6857	13.94
5	$4.02\ e^{-4}$	$4.14e^{-5}$	612.63	223.17	223829	21371	10.48	0.305	29.75	0.100	0.68838	13.91
6	$5.760e^{-4}$	$4.14e^{-5}$	611.64	214.73	201120	20821	9.659	0.307	29.87	0.1012	0.69276	13.87
7	$2.88e^{-4}$	$2.07e^{-5}$	613.82	231.82	246426	23127	10.66	0.290	27.91	0.0932	0.639	14.21
8	$2.88e^{-4}$	$6.21e^{-5}$	611.63	231.82	242368	20827	11.63	0.310	30.92	0.1050	0.7187	13.80

3.3 Effect of Reactor Diameter, Wall Heat Transfer Coefficient and Initial Solvent Concentration

Table 6 shows the effect of reactor diameter, the wall heat transfer coefficient, and the initial solvent concentration on the performance of LDPE reactor. With an increase in reactor diameter, the time taken to attain the peak temperature in the reactor increases (Fig. 12) and the monomer conversion decrease.

Effect of each parameter is studied independently keeping the values of other parameters constant at reference value (as given in Table D1 in Appendix D). Again the peak temperature remains almost constant with a change in values of these variables; however the location of peak temperature changes (Fig. 12). Polydispersity index increases with increasing the reactor diameter (Fig. 13 and Table 6). Polydispersity index is the ratio of weight-average molecular weight to the number-average molecular weight, which is a measure of the distribution of molecular mass in a given polymer sample. PDI also varies slightly with a variation in the wall heat transfer coefficient. The effect of presence of solvent is also tested on PDI. PDI shoots to a value of 37.84 when solvent concentration was neglected. Sr. No. 1 to 3 of Table 6 and Figs.14a - 14b show that the number-average molecular weight decreases marginally with an increase in diameter. But the weight-average molecular weight increases (with relatively high value) with an increase in the reactor diameter, which tends to increase the PDI. The number-average molecular weight increases substantially to a very high value in the first zone in the absence of solvent. The solvent plays an important role in controlling the temperature of the reactor and for terminating the chain to the solvent.

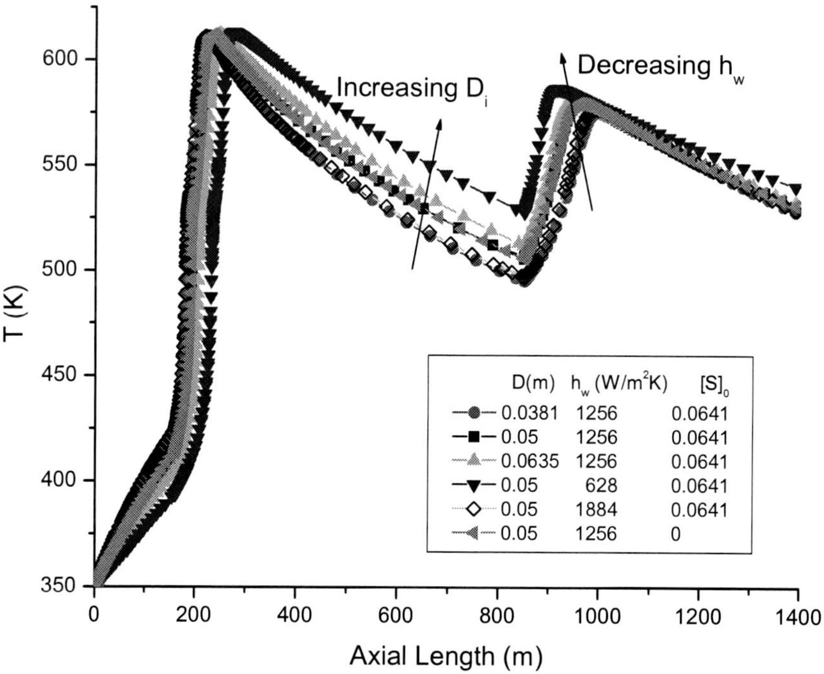

Figure 12. Influence of reactor diameter, wall heat transfer coefficient and initial solvent content on the temperature of the reactor.

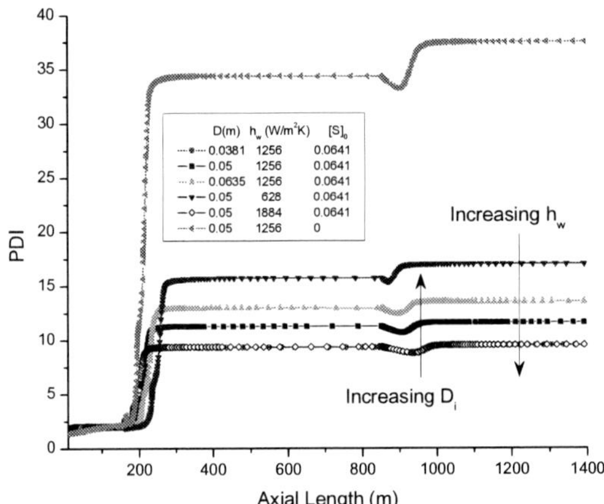

Figure 13. Influence of reactor diameter, wall heat transfer coefficient and initial solvent content on the polydispersity index.

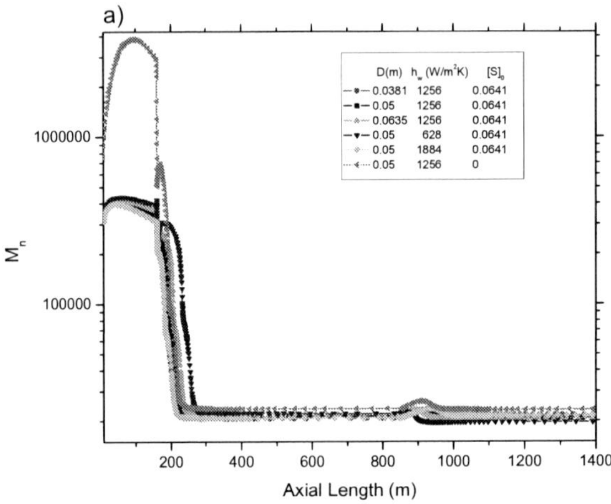

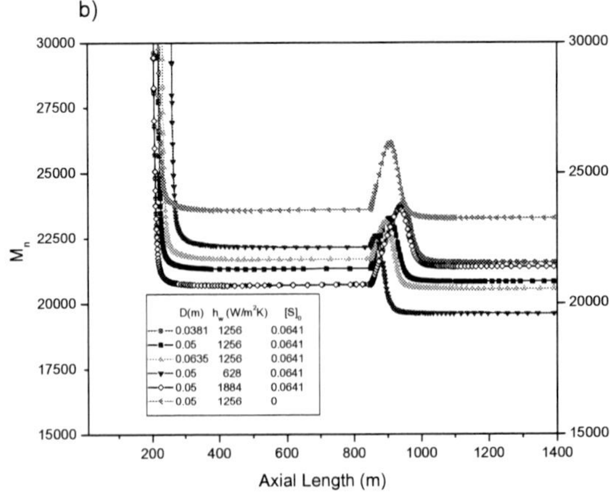

Figure 14. (a) Influence of reactor diameter, wall heat transfer coefficient and initial solvent content on the number-average molecular weight of polymer (b) Magnified part of Fig. (a) in the range of 15000-30000 M_n on y axes.

Table 6. Various properties and exit concentration of different species at different reactor diameter, the wall heat transfer coefficient and the initial solvent content

Sr. No	D_i (m)	h_w ($W/m^2.{}^0C$)	$[S]_0$ ($kmol/m^3$)	Max Temp. T_{MAX} (K)	Axial Position at Max Temp. L_a (m)	M_w (kg/kmol)	M_n (kg/kmol)	PDI	Xm (%) (Exit)	Exit Side chain concentration/10^3 C atoms			Unconverted Ethylene ($kmol/m^3$)
										Methyl Group (SCB)	Vinyl Group	Vinylidene Group	
1	0.038	1256	0.0641	610.95	218.77	205163	21558	9.5167	0.3188	30.065	0.1009	0.69214	13.79
2	0.05	1256	0.0641	611.63	231.82	242368	20827	11.636	0.3104	30.915	0.1050	0.71872	13.79
3	0.063	1256	0.0641	612.16	245.76	278176	20547	13.538	0.3089	31.213	0.1068	0.72945	13.74
4	0.05	628	0.0641	613.14	274.03	333787	19605	17.025	0.2975	32.297	0.1127	0.76675	13.83
5	0.05	1884	0.0641	610.99	218.01	204120	21386	9.5443	0.3162	30.259	0.1018	0.69818	13.84
6	0.05	1256	0	611.69	231.98	871875	23262	37.480	0.3110	30.951	0.1052	0.71984	13.78

An optimum choice of solvent is required in order to avoid the reactor instabilities and formation of side chains. In the absence of solvent, a relatively high value of Mn (=23, 262) is observed at the exit of reactor, which substantiates the above discussion. Figs. 15 - 16 show that the concentration of the side chain increases with increasing the reactor diameter, while it remains high at a lower value of the wall heat transfer coefficient. The side chain concentration is relatively independent of solvent concentration over the ranges covered in this study.

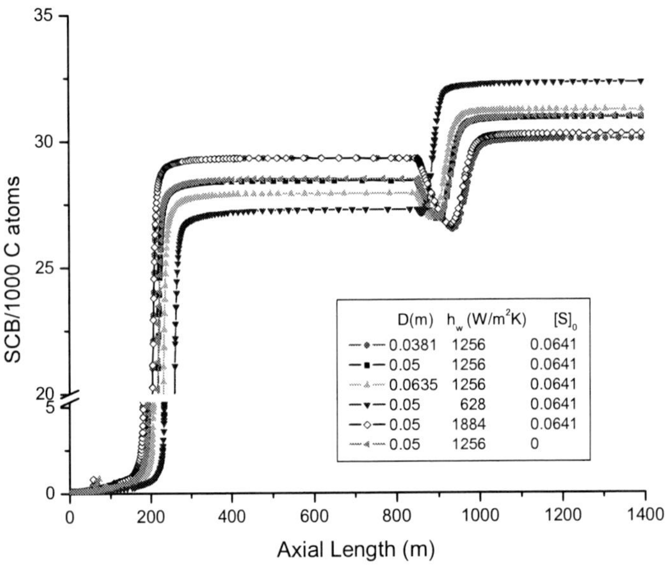

Figure 15. Influence of reactor diameter, wall heat transfer coefficient and initial solvent content on the short chain branching per 1000 C atoms along the reactor length.

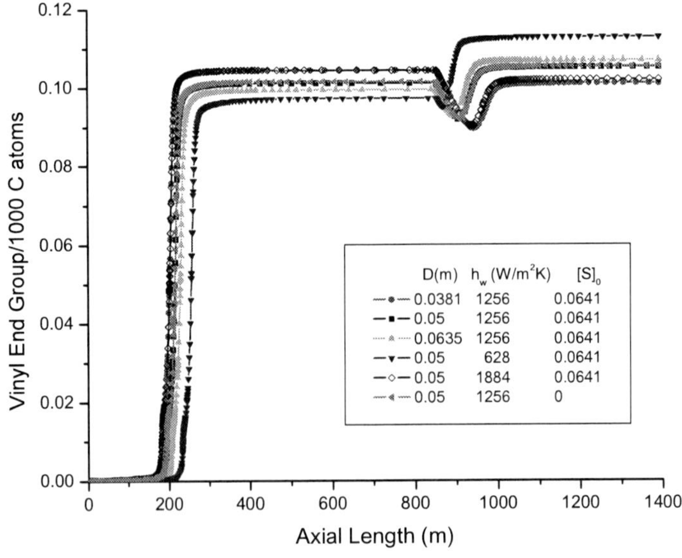

Figure 16. Influence of reactor diameter, wall heat transfer coefficient and initial solvent content on the vinyl end group per 1000 C atoms along the reactor length.

3.4 Effect of Initial Monomer Concentration and Initial Feed Velocity

The effect of initial concentration of monomer and the initial feed velocity on temperature, short chain branching and monomer conversion are shown in Figs. 17 – 19. With a low value of initial monomer concentration, the location at which the peak temperature is attained in the reactor is delayed (Fig. 17). The concentration of SCB and the monomer conversion is raised to a very high value when a relatively low monomer concentration was injected at the inlet of the reactor. With an increase in the initial monomer concentration (in the range considered in this study), the peak temperature increases while the concentration of side chain (SCB) and monomer conversion decrease as shown in Figs. 17 and 19. With an increase in the feed velocity, the monomer conversion increases while the concentration of SCB decreases. The reduction in SCB at relatively lower feed velocity is due to the lower value of maximum temperature attained in the reactor (see Figs. 17 and 18). With an increase in the feed velocity, the location of peak temperature is delayed as shown in Fig. 18.

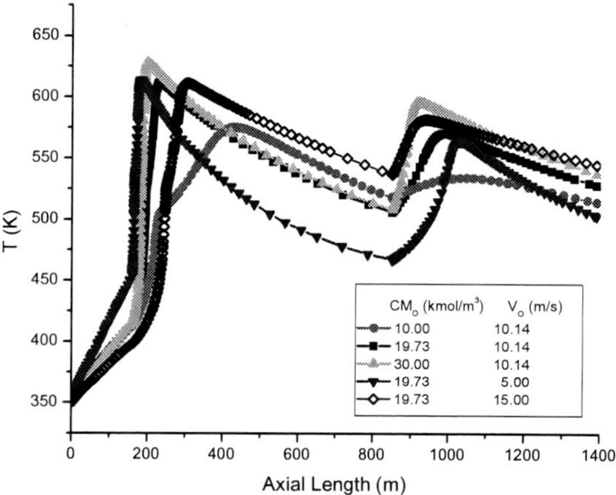

Figure 17. Influence of initial monomer flow rate and feed velocity on the temperature profile along the reactor length.

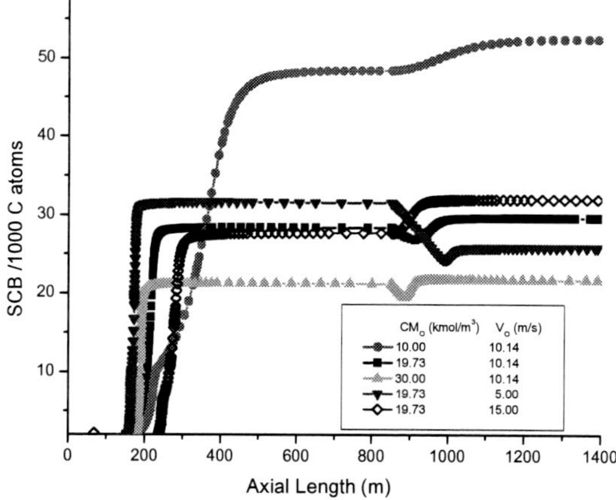

Figure 18. Influence of initial monomer flow rate and feed velocity on the short chain branching profile along the reactor length.

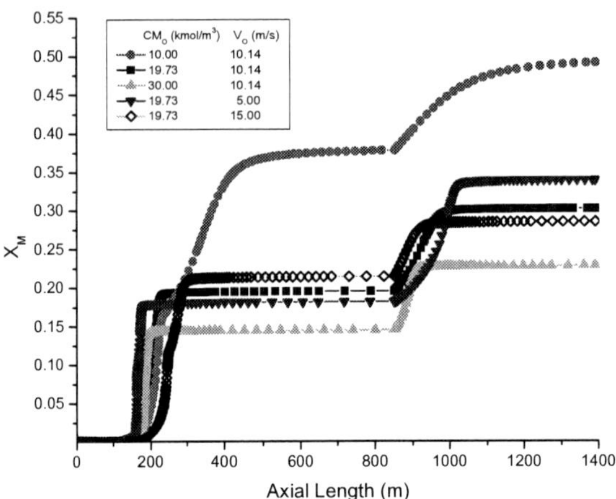

Figure 19. Influence of initial monomer flow rate and feed velocity on the monomer conversion profile along the reactor length.

Nomenclature

A	frequency Factor (1/s; m^3/kmol.s; m$^{3.3}$/kmol.s)
a_v	parameter used in the viscosity estimation
C_i	molar concentration of the ith component (kmol/m^3)
C_{Pi}	specific heat of reaction mixture in the ith zone (kJ/kg.K)
Di	inside diameter of the reactor (m)
D_{JI}	inner diameter of out jacket (m)
D_o	outer diameter of out jacket (m)
E	activation energy (kJ/kmol)
f	friction factor
F_i	mass flow rate for i^{th} component (kg/s)
f_m	initiator efficiency of the mth initiator
H_w	wall heat transfer coefficient (W/m^2.^{0}C)
I_i	i^{th} initiator
k	kinetic rate constants (1/s; m^3/kmol.s; m$^{3.3}$/kmol.s)
L	length of reactor (m)
L_a	reactor length at the maximum temperature (m)
M	monomer
M_e	methyl end group (SCB)
PE	poly ethylene
P_{in}	inlet pressure (MPa)
$R_{i(x)}$	growing macroradical with x monomeric units and i long chain branches
S	solvent (telogen)
t	wall thickness (m)
T	temperature of reaction mixture (K)
T_i	inlet temperature (K)
$T_{J,i}$	jacket fluid temperature on jacket side (K)
T_{MAX}	maximum temperature attained in the reactor (K)

SCB short chain branching
V_i vinyl end group
V_{id} vinylidene end group
$V_{J,i}$ volumetric flow rate of jacket side fluid (m^3/h)
v velocity of the reaction mixture (m/s)
X_M monomer conversion at any axial position
z axial length (m)
[i] concentration of ith species per 1000 C atoms

Greek Symbols

λ_{np} n,p order moments for the chain length distribution of macro radicals (kmol/m^3); n=0,1; p=0,1,2

μ_{np} n,p order moments for the chain length distribution of dead polymer molecules (kmol/m^3); n=0,1; p=0,1,2

Subscripts

b β-scission of a tertiary radical
bb backbiting or intramolecular chain transfer
b1 β-scission of a secondary radical
dm decomposition of ith peroxide (initiator); m=1,2
exit exit of the reactor
f reactor exit(m)
$I_{,m}$ m^{th} initiator, m=1,2
in reactor inlet
J Jacket side
M Monomer
O oxygen (initiator)
P propagation
S solvent
tc termination by combination
tdt thermal degradation
trp chain transfer to polymer
trs chain transfer to telogen or solvent

References

Agrawal, N., Rangaiah, G.P., Ray, A.K., Gupta, S. K. (2006). *Multi-objective Optimization of the Operation of an Industrial Low-Density Polyethylene Tubular Reactor using Genetic Algorithm and its Jumping Gene Adaptations,* Industiral and engineering chemistry research, 45, 3182-3199.

Agrawal, S., Han, C.D. (1975). *Analysis of the High-Pressure Polyethylene Tubular Reactor with Axial Mixing.* American Institute of Chemical Engineering Journal. 21, 449-465.

Asteasuain, M. Tonelli, S.M. Brandoline, A., Bandoni, J.A. (2001). *Dynamic Simulation and Optimization of Tubular Polymerization Reactors in gPROMS*, Computers and Chemical Engineering, 25, 509-515.

Babu, B. V., Angira, R. (2005). *Optimal Design of an Auto-thermal Ammonia Synthesis Reactor*, Computers & Chemical Engineering. 29, 1041-1045.

Babu, B. V., Chakole, P. G., Mubeen, J. H. S. (2005). *Multi-objective Differential Evolution (MODE) for Optimization of Adiabatic Styrene Reactor*. Chemical Engineering Science, 60, 4822-4837.

Babu, B. V., and Gujrathi, A. M. (2007). *Multi-Objective Differential Evolution (MODE) Algorithm for Multi-Objective Optimization: Parametric Study on Benchmark Test Problems*. Journal on Future Engineering & Technology, 3, 47-59.

Babu, B. V., Munawar, S. A. (2007). *Differential Evolution Strategies for Optimal Design of Shell-and-Tube Heat Exchangers*. Chemical Engineering Science, 62, 3720.

Babu, B. V., Sastry, K. K. N. (1999). *Estimation of heat transfer parameters in a trickle bed reactor using differential evolution and orthogonal collocation*. Computers and Chemical Engineering 23, 327–339.

Babu, B.V.,(2004) *Process Plant Simulation*, New York: Oxford University Press.

Brandoline, Capiati, N.J., Farber, J.N., Valles, E.M. (1988). *Mathematical Model for High-Pressure Tubular Reactor for Ethylene Polymerization*. Industrial and Engineering Chemistry Research, 27, 784-790.

Brandoline, Lucanza, M.H., Urgin, P.E., Capiati, N.J. (1996). *High-Pressure Polymerization of Ethylene. An Improved Mathematical Model for Industrial Tubular Reactors*, Polymer Reaction Engineering, 4, 193-241.

Brandoline, Valles, E. M., Farber, J. N. (1991). *High-Pressure Tubular Reactors for Ethylene Polymerization Optimization Aspects*. Polymer Engineering Science, 31, 381-390.

Chakraborti, N., Kumar, A. (2003). *The optimal scheduling of a reversing strip mill: studies using multipopulation genetic algorithms and differential evolution*, Materials and manufacturing Processes, 18, 433-445.

Chen, C.H., Vermeychuk, J.G., Howell, J. A., Ehrlich, P. (1976). *Computer Model for Tubular High-Pressure Polyethylene Reactors*. American Institute of Chemical Engineering Journal. 22, 463-471.

Dhib, R. Al-Nidawy, N. (2002). *Modeling of free radical polymerization of ethylene using difunctional initiators*, Chemical Engineering Science, 57, 2735-2746.

Goto, S., Yamamoto, K.S., Furui, S., Sugimoto, M. (1981). *Computer Model for Commercial High-Pressure Polyethylene Reactor Based on Elementary Reaction Rates Obtained Experimentally*. Journal of Applied Polymer Science, 36, 21-40.

Gujarathi, A.M., Babu, B.V. (2009a). *Improved Multi-Objective Differential Evolution (MODE) Approach for Purified Terephthalic Acid (PTA) Oxidation Process*, Materials and Manufacturing Processes. 24, 303-319, 2009.

Gujarathi, A.M., Babu, B.V. (2009b). *Hybrid Multi-objective Differential Evolution (H-MODE) for optimization of Polyethylene Terephthalate (PET) reactor. International* Journal of Bio-inspired Computation. Accepted (In Press).

Gujarathi, A.M., Babu, B.V. (2009c). *Optimization of Adiabatic Styrene Reactor: A Hybrid Multi-Objective Differential Evolution (H–MODE) Approach*. Industrial & Engineering Chemistry Research. Accepted (In Press). Available online at http://pubs.acs.org/doi/pdf/10.1021/ie901074k

Gujarathi, A.M., Babu, B.V. (2009d) *Multi-objective Optimization of Industrial Styrene Reactor: Adiabatic and Pseudo-isothermal Operation*. Chemical Engineering Science. Accepted (In Press)

Gupta, S. K., Kumar, A., Krishnamurthy, M. V. G. (1985). *Simulation of Tubular Low-Density Polyethylene,* Polymer Engineering and Science, 25, 37-47.

Kalyon, D.M., Chiou, Y.N., Kovenklioglu, S., Bouaffar, A. (1994). *High Pressure Polymerization of Ethylene and Rheological Behavior of Polyethylene Product.* Polymer Engineering and Science, 34, 804-814.

Katz, S., Saidel, G.M. (1967). *Moments of the Size Distribution in Radical Polymerization,* AIChE Journal, 13, 319-326.

Kim, D. M., Iedema, P. D. (2008). *Modeling of branching density and branching distribution in low density polyethylene polymerization.* Chemical Engineering Science, 63, 2035-2046.

Kiparissides, C., Verros, G., MacGregor, J. F. (1993). *Mathematical Modeling, Optimization, and Quality Control of High Pressure Ethylene Polymerization Reactors,* Polymer Reviews, C33, 437-527.

Lacunza, M. H.; Ugrin, P. E.; Brandolin, A.; Capiati, N. J. (1998). *Heat Transfer Coefficient in a High-Pressure Tubular Reactor for Ethylene Polymerization.* Polymer Engineering Science, 38, 992-1013

Lee, H. J., Yeo, Y. K., Chang J. Y. (2000). *Modeling of industrial high pressure autoclave polyethylene reactor including decomposition phenomena,* Korean Journal of Chemical Engineering, 17, 223-229.

Lee, M. H., Han, C. , Chang, K. S. .(1999). *Dynamic Optimization of a Continuous Polymer Reactor Using a Modified Differential Evolution Algorithm, Industrial and Engineering* Chemistry Research, 38, 4825–4831.

Price, K. V., Storn, R. (1997). *Differential evolution - a simple evolution strategy for fast optimization.* Dr. Dobb's Journal, 22, 18-22.

Shampine, L.F., Reichelt, M. (1997). *The matlab ODE suite,* SIAM Journal in Scientific Computing, 18, 1-22.

Shirodkar, P.P., Taien, G.O. (1986). *A mathematical model for the production of low density polyethylene in a tubular reactor.* Chemical Engineering Science, 41, 1031-1038.

Stumberger, G., Dolinar, D. , Pahner, U., Hameyer, K. (2000). *Optimization of radial active magnetic bearings using the finite element technique and the differential evolution algorithm,* IEEE Transactions on Magnetics 36, 1009-1013.

Yao, F. Z., Lohi, A., Upreti, S. R., Dhib, R. (2004). *Modeling, Simulation and Optimal Control of Ethylene Polymerization in Non-Isothermal, High-Pressure Tubular Reactors. International* Journal of Chemical Reactor Engineering, 2, 1-25.

APPENDIX

Reaction scheme involved in the LDPE reactor

Oxygen Initiation $O_2 + M \xrightarrow{k_o} 2R_1(O)$ (A1)

Peroxide Initiation $I_m \xrightarrow{f_m k_{dm}} 2R_1(O)$ (for m = 1,2) (A2)

Propagation $R_i(x) + M \xrightarrow{k_p} R_i(x+1)$ (A3)

Termination by combination $R_i(x) + R_j(y) \xrightarrow{k_{tc}} P_{i+j-1}(x+y)$ (A4)

Termination by Thermal degradation $R_i(x+1) + M \xrightarrow{k_{tdt}} P_i(x) + R_1(O)$ (A5)

Chain transfer to telogen $R_i(x) + S \xrightarrow{k_{trs}} P_i(x) + R_1(O)$ (A6)

Chain transfer to polymer $R_i(x) + P_j(y) \xrightarrow{k_{trp}} P_i(x) + R_{j+1}(y)$ (A7)

Intramolecular chain transfer $R_i(x) \xrightarrow{k_{bb}} R_i(x)$ (A8)

(short-chain branching)

β-scission of secondary radical $R_i(x+1) \xrightarrow{k_{b1}} P_i(x) + R_1(O)$ (A9)

(formation of vinyl group)

β-scission of territory radical $R_i(x+1) \xrightarrow{k_b} P_i(x) + R_1(O)$ (A10)

(formation of vinylidene group)

Model Equations

Continuity equation

$$\frac{dv}{dz} = -\frac{v}{\rho}\left(\frac{d\rho}{dz}\right) \tag{A11}$$

Mass balance Model Equations for components j

$$v\frac{d\left[C_j\right]}{dz} = -\left(r_j + [C_j]\frac{dv}{dz}\right) \tag{A12}$$

Where j=3, 4,..11 for I_1, I_2, O_2, M, S, M_e, V_i, and V_{id}, $R_{i(0)}$ respectively
Reactor temperature: Heating and reaction zones

$$\rho C_p v\frac{dT}{dz} = -\left\{\frac{4U\left(T - T_j\right)}{1000D_i} + k_p C_M \lambda_{00}\left(\Delta H\right)\right\} \tag{A13}$$

Frictional pressure drop

$$\frac{dP}{dz} = -10^{-6}\left\{\frac{2f_r\rho v^2}{D_i} + \rho v\frac{dv}{dz}\right\} \tag{A14}$$

Moment equations related to dead and live polymer

$$v\frac{d\lfloor\lambda_{np}\rfloor}{dz} = -\left(r_{\lambda_{np}} + \left[\lambda_{np}\right]\frac{dv}{dz}\right) \quad \text{(for n = 0,1... and p = 0,1,2,...)} \tag{A15}$$

Where

$$r_{\lambda_{np}} = -\begin{pmatrix} 2k_o C_M C_{O_2}^{1.1}\delta_{p0} + 2\sum_{j=1}^{2} f_j k_{d_j}\delta_{p0} + pk_p C_M \lambda_{n,p-1} - k_{tc}\lambda_{00}\lambda_{np} + k_{tdt}\left(\lambda_{00}\delta_{p0} - \lambda_{np}\right) \\ + k_{trs}C_S\left(\lambda_{00}\delta_{p0} - \lambda_{np}\right) + k_{trp}\left(\lambda_{00}\sum_{j=0}^{n}\left({}^nC_j\right)\mu_{j,p+1} - \lambda_{np}\mu_{01}\right) \end{pmatrix} \tag{A16}$$

$$v\frac{d\lfloor\mu_{np}\rfloor}{dz} = -\left(r_{\mu_{np}} + \left[\mu_{np}\right]\frac{dv}{dz}\right) \quad \text{(for n = 0,1... and p = 0,1,2,...)} \tag{A17}$$

Where,

$$r_{\mu_{np}} = -\begin{pmatrix} \frac{ktc}{2}\sum_{j=0}^{n}\sum_{k=0}^{j}\sum_{i=0}^{p}\left({}^nC_j\right)\left({}^jC_k\right)\left({}^pC_i\right)\left(-1\right)^{n-j}\lambda_{ki}\lambda_{j-k,p-i} \\ + k_{tdt}\lambda_{np} + k_{trs}C_S\lambda_{np} + k_{trp}\left(\lambda_{np}\mu_{01} - \lambda_{00}\mu_{n,p+1}\right) \end{pmatrix} \tag{A18}$$

$$\mu_{03} = \mu_{00}\left(\frac{\mu_{02}}{\mu_{01}}\right)^3 \tag{A19}$$

$$\mu_{13} = \mu_{10}\left(\frac{\mu_{12}}{\mu_{11}}\right)^3 \tag{A20}$$

$$X_M = 1 - \frac{v}{v_{in}}\left(\frac{C_M}{C_{M_{in}}}\right) \tag{A21}$$

$$M_n = 28\left(\frac{\mu_{01} + \lambda_{01}}{\mu_{00} + \lambda_{00}}\right) \tag{A22}$$

$$M_W = 28\left(\frac{\mu_{02} + \lambda_{02}}{\mu_{01} + \lambda_{01}}\right) \tag{A23}$$

$$PDI = \frac{M_W}{M_n} \tag{A24}$$

$$\text{Methyl Group (SCB) per } 10^3 \text{ C atoms} = \frac{500 C_{M_{exit}}}{\mu_{01} + \lambda_{01}} \tag{A25}$$

$$\text{Vinyl Group per } 10^3 \text{ C atoms} = \frac{500 C_{Vi_{exit}}}{\mu_{01} + \lambda_{01}} \tag{A26}$$

$$\text{Vinylidene Group per } 10^3 \text{ C atoms} = \frac{500 C_{Vid_{exit}}}{\mu_{01} + \lambda_{01}} \tag{A27}$$

$$k = A \exp\left(-\frac{E + 10^3 P\Delta V}{RT} \right) \tag{A28}$$

$f = 0.316/(\mathrm{Re})^{1/4}$ from Gupta et al. [7] (Case A) (A29a)

$f = \frac{16}{\mathrm{Re}};\quad \mathrm{Re} < 2100$ from Kiparissides et al. [13] (Case B) (A29b)

$(\mathrm{Re})^{1/2} = 4 \log\left(\mathrm{Re}\, f^{1/2}\right) - 0.4;\quad 2.1 \times 10^3 < \mathrm{Re} < 5 \times 10^6$ (Case B) (A29c)

In: Handbook of Research on Chemoinformatics ... ISBN: 978-1-62100-998-6
Editor: A.K. Haghi

Chapter 7

HYDROTHERMAL STABILITY OF YTTRIA AND NEODYMIA CO-STABILISED ZRO2 AND ZRO2 BASED COMPOSITES

Sedigheh Salehi, Akhilesh Kumar Swarnakar, Omer Van der Biest and Jef Vleugels*

Department of Metallurgy and Materials Engineering, K.U.Leuven, Kasteelpark Arenberg 44, B-3001 Leuven, Belgium

1. INTRODUCTION

One of the drawbacks of tetragonal ZrO_2 polycrystalline ceramics (TZP) is the tetragonal to monoclinic transformation due to aging at low temperature, known as Low temperature degradation (LTD), especially in aqueous environments [1-4]. In assessing LTD, two major items are important, i.e. the rate of tetragonal to monoclinic ZrO_2 phase transformation on the material surface and the rate of LTD extension from the surface into the bulk of the material, ultimately resulting in material disintegration and fracture. The most important factors enhancing the susceptibility of the material to LTD are humidity and increased t-ZrO_2 phase transformability, either induced by a decreased stabiliser content or an increased grain size[2]. Several mechanisms have been claimed in literature [1,5,6]. The earlier models explaining the LTD behaviour in terms of the reaction between water and yttria forming $Y(OH)_3$ during aging have been rejected and the role of internal stresses associated with the diffusion of water into the ZrO_2 lattice has been demonstrated [7,8]. LTD is now considered to be a relaxation process of the internally strained lattice by thermally activated oxygen vacancy diffusion [1].

Amongst the methods to improve the resistance of ZrO_2-based materials to LTD are, reduction of the ZrO_2 grain size, surface coating of the component with a thin layer of stable c-ZrO_2, addition of alumina particles, the presence of an amorphous grain boundary phase,

* Corresponding author: MTM, Kasteelpark Arenberg 44 - bus 2450, 3001 Heverlee, Belgium, Phone: +32 16 321244, e-mail: Jozef.Vleugels@mtm.kuleuven.be

increasing the stabiliser content and using alternative stabilisers such as CeO_2. Although no satisfactory solution has been found, an increased stability of the t-ZrO_2 phase remains the most favourable approach, resulting however in a decreased t-ZrO_2 phase transformability and concomitant transformation toughening. [2,9]

Experimental work on Y_2O_3-Nd_2O_3 double stabilised ZrO_2 ceramics revealed the possibility to establish (Y,Nd)-TZP's with excellent fracture toughness [10,11]. In this chapter, the hydrothermal stability of (Y,Nd)-TZP and their composites is compared with conventional Y-TZP and Y-TZP based composites.

2. Experimental Procedure

Neodymia and yttria stabilisers were applied to a monoclinic ZrO_2 nanopowder to obtain 2 and 3 mol % Y_2O_3 (grades 2Y and 3Y) and 1 mol % Y_2O_3 + 1-4 mol % Nd_2O_3-coated powder (grades 1Y-1Nd, 1Y-2Nd, 1Y-3Nd and 1Y-4Nd) [12]. As a reference powder, a 3 mol % Y_2O_3-coprecipitated powder (grade TZ-3Y, Tosoh, Tokyo, Japan) was used. The powders were densified into 5 mm thick 30 mm diameter discs by means of PECS for 3 min at 1450°C and 62 MPa.

The microstructure, Vickers hardness, indentation toughness, elastic modulus, and Low temperature degradation (LTD) tests were investigated.

Hydrothermal aging tests were performed at 200°C under a saturated H_2O pressure of 1.55 MPa. For this purpose, rectangular bars ($20 \times 5 \times 2$ mm) were polished and inserted in a stainless steel autoclave. The autoclave was placed in a salt bath to establish an internal temperature of 200°C, as monitored by a thermocouple in the autoclave. About 12 minutes are needed to reach a constant temperature of 200°C. The reported LTD testing time only encompasses the time at 200°C. The samples were positioned in a porous polyethylene basket above the water level in the autoclave. The amount of m-ZrO_2 on the exposed sample surface was measured by X-ray diffraction. The transformability during hydrothermal aging is the % m-ZrO_2 difference between the steam-treated and pristine polished surface.

3. Results and Discussion

3.1 Hydrothermal Aging of ZrO_2 Ceramics

The evolution of the m-ZrO_2 content on the surface of the investigated ZrO_2 grades during hydrothermal testing is summarized in Figure 1. In all material grades, the t-ZrO_2 phase was stabilised after sintering. The amount of m-ZrO_2 on the polished surface of the 1Y-1Nd and 1Y-2Nd grades is higher than for the 2Y grade (7, 5 and 2 vol % respectively). It is not impossible that this small amount of m-ZrO_2 is caused during final polishing of these highly transformable grades.

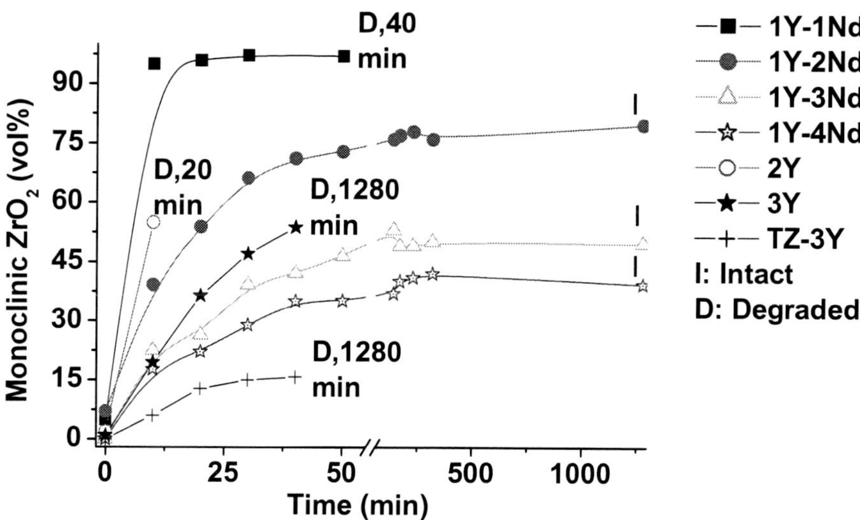

Figure 1. Evolution of the m-ZrO2 content on the surface of the initially polished material grades as a function of the residence time in 1.5 MPa water vapour pressure at 200°C.

The 2 mol % Y_2O_3 coated powder based material (2Y) degraded completely after 20 minutes, whereas the 1Y-1Nd ceramic degraded after 40 min, although about 95% of the t-ZrO_2 phase already transformed after only 10 min. The 1 mol % Y_2O_3 co-stabilised grades with 2, 3 and 4 mol % Nd_2O_3 remained intact after testing for 1280 min, whereas the 3 mol % yttria coated (3Y) and co-precipitated (TZ-3Y) grades degraded after 1280 min. The sample morphology after testing of the 3Y, TZ-3Y and 1Y-2Nd grades are shown in Figure 2. Since the dimensions of all samples were exactly the same, the fact that the (1Y,2Nd)-TZP grade was still intact after 1280 min clearly indicates a higher resistance against hydrothermal degradation.

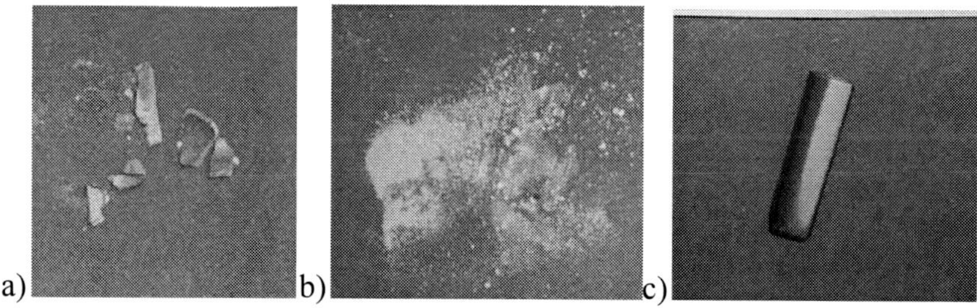

Figure 2. Sample morphology after 1280 min of hydrothermal testing. 3 mol % Y_2O_3 coated (a), 3 mol % Y_2O_3 co-precipitated (b) and 1 mol % Y_2O_3 + 2 mol % Nd_2O_3 coated (c) grades.

The hydrothermal stability of the yttria coated powder based materials increases with increasing yttria content, confirming literature data stating that a ZrO_2 with higher transformability is more susceptible to LTD [1]. Although the hydrothermal transformability of the 3 mol % yttria co-precipitated grade is lower than for the 3 mol % yttria coated grade, both materials completely degraded after 1280 min as shown in Figure 2.

Accordingly, the hydrothermal stability of the co-stabiliser grades was also found to increase with increasing Nd_2O_3 co-stabiliser content. In contrast to the yttria-stabilised ceramic grades, the (1Y,2-4Nd)-TZP grades did not disintegrate even after 1280 min in steam. So, although the mixed stabiliser ceramics are also prone to hydrothermally induced phase transformation, their resistance to fracture and complete degradation is higher than for the yttria-stabilised material with the same overall stabiliser content.

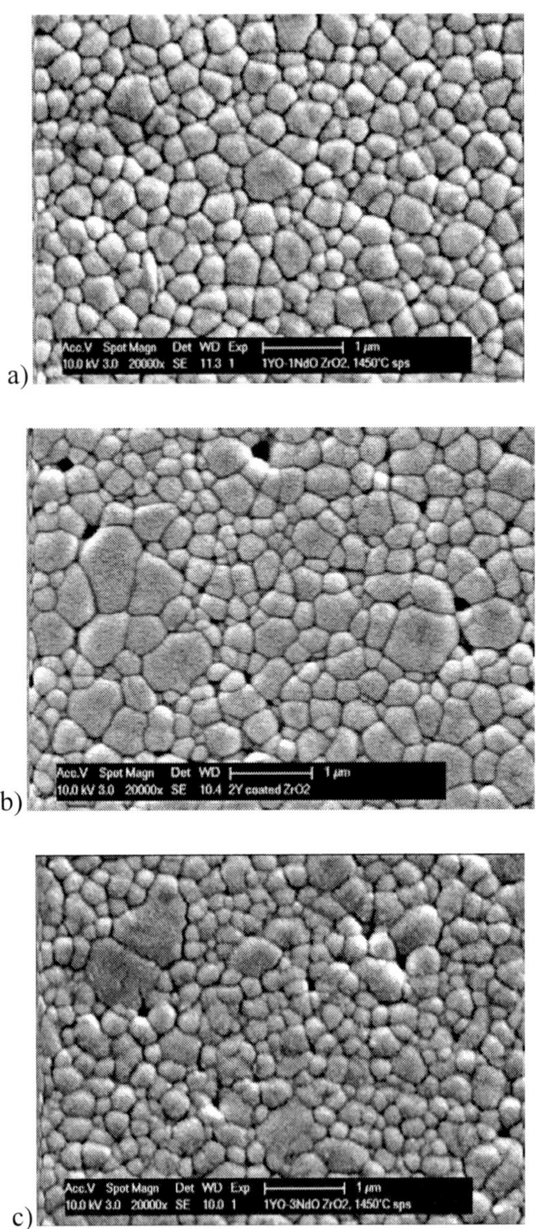

Figure 3. Secondary electron micrographs of co-stabilised (1Y,Nd)-ZrO_2 ceramics with 1 (a) and 3 (c) mol % Nd_2O_3, together with 2 mol % Y_2O_3 coated ZrO_2 powder based ceramics (b).

The mechanical properties of the Y- and (Y,Nd)-TZPs are summarized in Table 1. The transformability and concomitant toughness of the yttria-stabilised ceramics is modest due to the very short dwell time of 3 min at 1450°C during PECS, limiting ZrO_2 grain growth. The transformability and toughness of the double stabiliser systems is substantially higher. The average grain size of the (1Y,1Nd)-TZP however is even smaller than for the 2Y-TZP grade (Figure 3a and b). The transformability and toughness decreases with increasing Nd_2O_3 co-stabiliser content, what can be explained by the formation of a bimodal tetragonal and cubic microstructure at higher Nd_2O_3 levels (Figure 3c). The hardness and stiffness of the ZrO_2 grades are comparable.

Table 1. Mechanical properties of the different ZrO_2 grades, PECS at 1450°C for 3 min at 62 MPa

Stabiliser (mol %)	Density (g/cm^3)	E (GPa)	HV (GPa)	K_{Ic} (MPa.m$^{1/2}$)	$^{i}t_{tra}$ (%)
2Y	5.99	170	12.59 ± 0.04	4.1 ± 0.1	5.2
3Y	5.98	185	12.97 ± 0.13	4.0 ± 0.2	4.0
^{ii}TZ-3Y^{2}	6.02	179	12.40 ± 0.13	3.4 ± 0.1	2.0
1Y-1Nd	5.97	179	10.76 ± 0.48	7.4 ± 0.5	63.9
1Y-2Nd	6.03	180	11.17 ± 0.47	5.9 ± 0.2	51.1
1Y-3Nd	6.17	193	12.29 ± 0.14	4.5 ± 0.1	36.5
1Y4Nd	6.13	197	12.50 ± 0.08	4.4 ± 0.2	0.0

^{i} tetragonal ZrO_2 transformability - ^{ii} 3 mol % Y_2O_3 co-precipitated starting powder

All investigated ZrO_2 grades are fully dense since no residual porosity could be observed on polished cross-sections by means of SEM. The evolution of the density in Table 1 should therefore be attributed to the stabiliser additions.

The XRD patterns of the main tetragonal peak around 30.25° 2θ of the 2Y, 3 Y and 1Y-xNd grades are compared in Figure 4, showing a shift to lower 2θ values, i.e. larger lattice parameters and lower tetragonality with increasing Nd_2O_3 content. Since the substitution of Y by Nd has no effect on the amount of oxygen vacancies in the lattice, the larger lattice parameters of the 1Y-1Nd and 1Y-2Nd grades have to be attributed to the larger ionic radius of Nd^{3+}, 0.1109 nm, compared to Y^{3+}, 0.1019 nm. The peak broadening of the 1Y-3Nd grade can be attributed to the partially overlapping t- and c-ZrO_2 peak. The experiments indicate that the wider t-ZrO_2 lattice makes the material less susceptible to internal straining by water diffusion and accompanying LTD by relaxation. Although Y-TZP as well as Ce-TZP's with a CeO_2 content < 10 mol % exhibit aging-induced transformation, the addition of both CeO_2 and Y_2O_3 was similarly found to be beneficial to obtain hydrothermally stable (Ce,Y)-TZP's. [13]

Dilatometer data on the 3Y, 1Y-1Nd and 1Y-2Nd grades are compared in Figure 5. The 3Y grade shows a volume expansion around 200°C during heating, corresponding with a t to m phase transformation that should be attributed to LTD. The m-ZrO_2 phase is converted to t-ZrO_2 when heating above 500°C. During cooling, the martensitic transformation is clearly observed with an A_s temperature around 450°C. Despite the high transformability of the t-ZrO_2 phase (see Table 1), no phase transformations were observed for the 1Y-1Nd and 1Y-2Nd grades during heating to 800°C and subsequent cooling to room temperature, showing that the martensitic transformation in the double stabiliser grades is not thermally activated within the investigated temperature region. The fact that the transformation can only be mechanically activated makes these materials more resistant against hydrothermal degradation compared to the yttria stabilised system, which can also be thermally activated. The transformation behaviour of the double stabilised system is comparable with that of 12Ce-TZP that has an M_s temperature well below room temperature and is not susceptible to LTD.

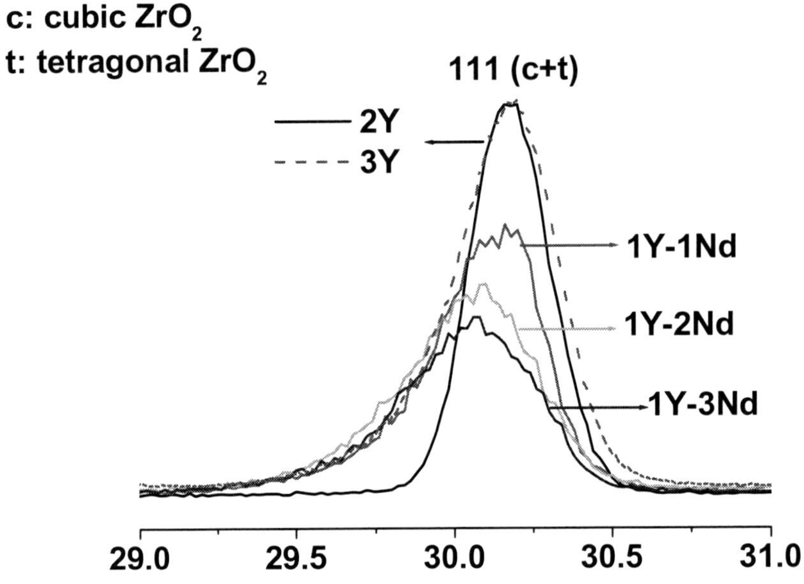

Figure 4. XRD patterns of the 2Y and 1Y-xNd grades.

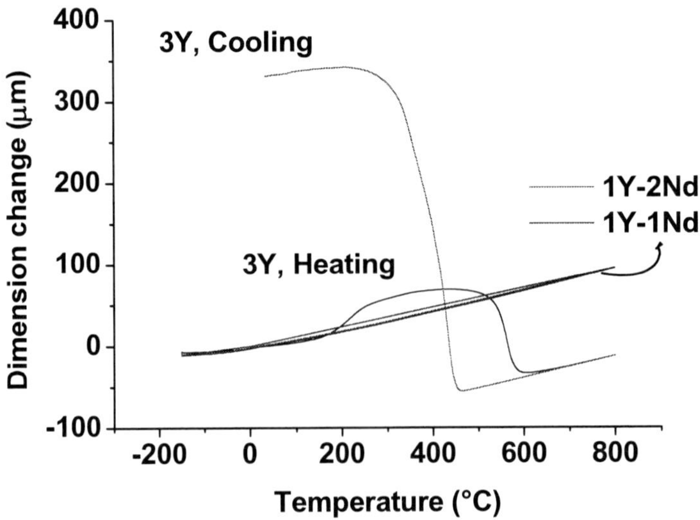

Figure 5. Dilatometer data of the 3Y, 1Y-1Nd and 1Y-2Nd ceramics.

The higher stability of the Y-Nd co-stabilised systems compared to that of yttria stabilised systems opens up the possibility to increase the reliability of hydrothermally stable Y-TZP ceramics for example for biomedical applications by the partial replacement of the stabiliser, without drastic changes in the fracture toughness. More work however is needed to fully reveal the exact contribution of the Nd_2O_3 addition.

3.2 Hydrothermal Aging of ZrO2 Based Composites

3.2.1 Influence of the Secondary Phase Content

To assess the influence of the NbC amount on the hydrothermal stability of ZrO_2-NbC composites, 2 mol % yttria stabilised ZrO_2-NbC composites with 40, 60 and 70 vol % NbC were exposed in an interrupted hydrothermal stability test. The amount of m-ZrO_2 phase, i.e., the fraction of m-ZrO_2 in the total ZrO_2 phase, measured on the initially polished surface of the samples is plotted as a function of the aging time at 200°C in Figure 6. The microstructure and mechanical properties were reported in Section 3.2.1.

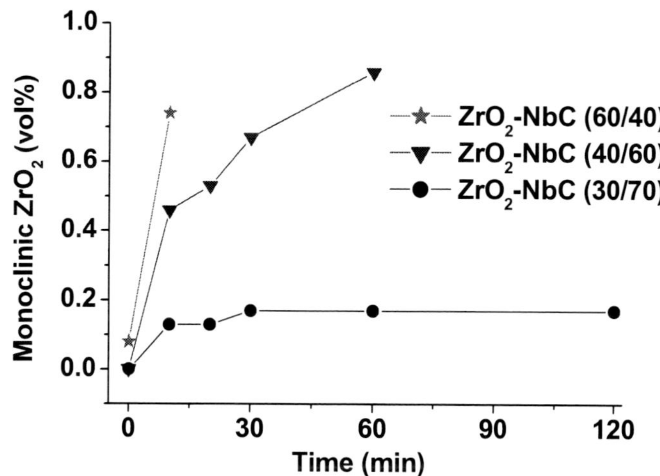

Figure 6. Evolution of the m-ZrO_2 fraction of ZrO_2 phase on the surface of initially polished of ZrO_2-NbC composites with 40, 60 and 70 vol % NbC during aging in saturated H_2O vapour at 200°C.

The ZrO_2-NbC (60/40) composite degraded after only 10 minutes at 200°C. The composite with 60 vol % NbC broke into two pieces after 1 h of testing whereas the ceramic with 70 vol % NbC remained intact with a raise of the surface m-ZrO_2 content up to 19 % after about 7 h of testing.

Although the XRD data only provide information on the top 10 μm of the material, it is clear that the resistance to LTD increases with increasing NbC content. Although LTD is strongly reduced in the composite with 70 vol % NbC, despite the residual porosity, it could not be completely inhibited.

During LTD, the accumulation of tensile strain due to OH diffusion into the t-ZrO_2 grains initiates the transformation to m-ZrO_2 and a subsequent transformation of neighbouring grains [14,15]. The t-ZrO_2 lattice expansion due to OH adsorption can be partially restricted by the presence of NbC grains with a higher E modulus, similar to the influence of Al_2O_3 in zirconia toughened alumina and alumina doped zirconia [3,14] and the addition of other secondary phases such as MgO, mullite or WC to a ZrO_2 matrix [16]. Moreover, the progressing transformation of neighbouring grains can be suppressed by the interference of non-transforming NbC grains. These mechanisms can explain the increasing resistance to LTD with increasing NbC content in the ZrO_2-NbC composites.

It was also shown in Section 3.2.1 that the position of the tetragonal peak moves towards a lower 2θ angle as a function of the NbC content (Figure 7, Section 3.2.1), implying an increased lattice spacing. Shifting of the tetragonal ZrO_2 peak towards a lower 2θ in the XRD pattern could be interpreted as an "overstabilisation" of tetragonal ZrO_2 or of partially remaining as cubic ZrO_2 at room temperature.

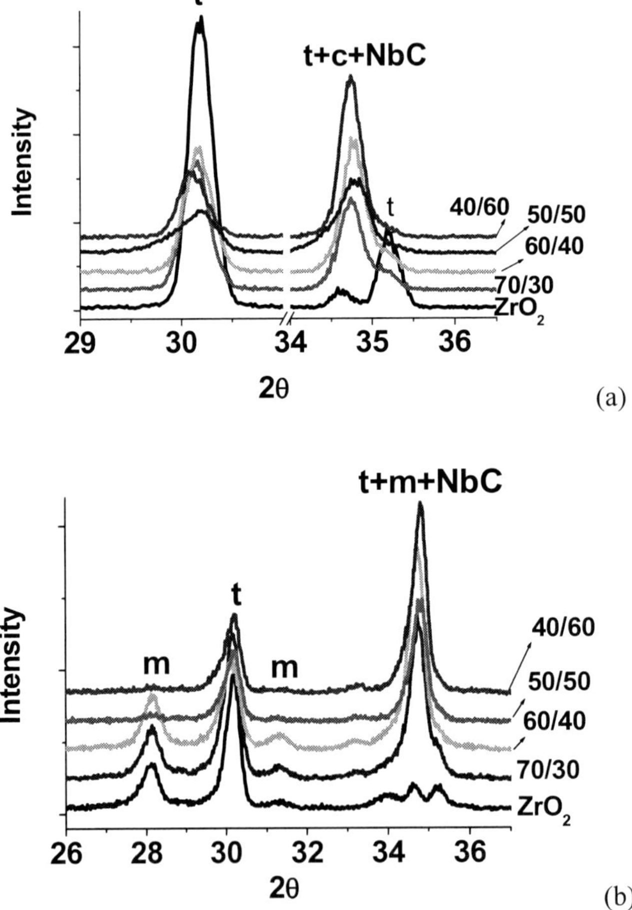

Figure 7. XRD patterns of polished surfaces of 2 mol % yttria stabilised ZrO_2 and ZrO_2-NbC composites after hot pressing at 1450°C for 1 h (a) and after 2 years stored at room temperature (b). t = tetragonal, m = monoclinic and c = cubic ZrO_2.

After two years of storage at room temperature and ambient pressure, ZrO_2-NbC composites with less than 50 vol % NbC were found to crack. The XRD patterns of polished surfaces of samples before and after two years are shown in Figure 7. Although low temperature degradation (LTD) of ZrO_2 is reported to be enhanced at elevated temperature in humid environment [13], these results prove that it can also occur at room temperature [14]. This clearly illustrates that low temperature degradation is of major concern when dealing with ZrO_2 ceramics, especially in humid environments.

3.2.2 Influence of Stabiliser and Secondary Phase Type and Content

Hydrothermal testing was performed on 2 mol % Y_2O_3 and 1 mol % Y_2O_3+1 or 2 mol % Nd_2O_3 stabilised ZrO_2-TiCN (60/40) nanocomposites (grade nano-T), PECS at 1450°C. The 2 mol % Y_2O_3 stabilised composite complete degraded after 10 min exposure in 1.5 MPa water vapour pressure at 200°C while the 1 mol yttria + 1 mol neodymia co-stabilised composite degraded only after 30 minutes. The (1Y,2Nd) nanocomposite broke after 120 minutes hydrothermal testing as illustrated in Figure 8.

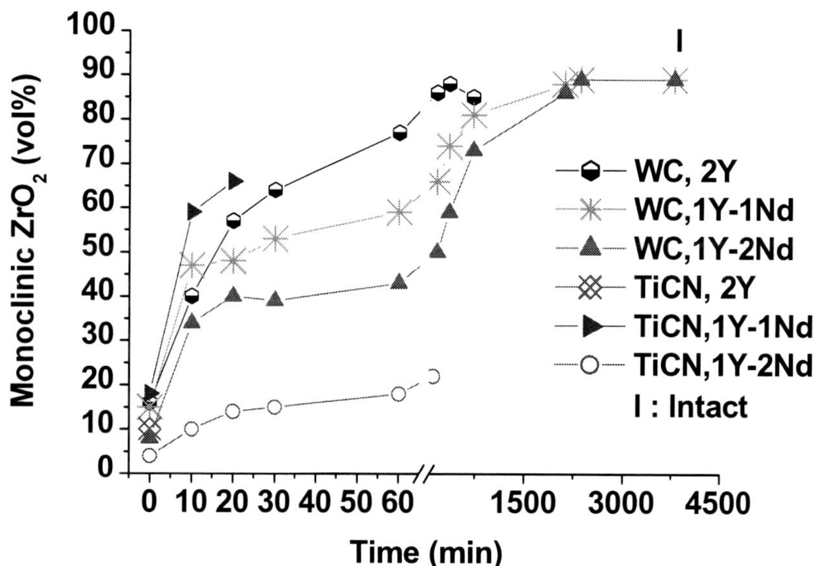

Figure 8. Evolution of the m-ZrO_2 fraction of ZrO_2 phase on the surface of initially polished of 2 mol % Y_2O_3 (2Y) and 1 mol % Y_2O_3 + 1 or 2 mol % Nd_2O_3 (1Y-1 or 2Nd) stabilised ZrO_2 based composites with 40 vol % WC or TiCN secondary phases.

Although the (1Y,1Nd) stabilised composite has the same overall amount of stabiliser as the 2 mol % yttria stabilised material, the hydrothermal stability of the double stabiliser ZrO_2-TiCN composite is about 3 times higher. Moreover, the hydrothermal stability increases with increasing Nd_2O_3 co-stabiliser content.

The results of the hydrothermal stability testing, performed on 2 mol % yttria (2Y) and 1 mol % yttria + 1 mol % neodymia (1Y,1Nd) stabilised ZrO_2-WC (60/40) composites PECS at 1450°C, are also summarised in Figure 8. The 2 mol % yttria stabilised composite completely degraded after 2000 minutes in 1.5 MPa saturated water vapour at 200°C, while the 1Y-1Nd and 1Y-2Nd stabilised ZrO_2-WC composites remained intact even after 4000 minutes.

In general, mechanisms, which increase hydrothermal stability such as reduction of the ZrO2 grain size, increasing the stabiliser content, using alternative stabilisers such as CeO2, etc. [2,9] decrease t-ZrO2 phase transformability and consequently fracture toughness. Beside the higher hydrothermal stability, the (1Y,1Nd) stabilised ZrO_2-WC composites also have a higher fracture toughness than the 2Y stabilised ZrO_2-WC [17]. When using a yttria-neodymia double stabiliser system, it is possible to increase both fracture toughness and hydrothermal stability.

The reason for the superior hydrothermal stability of double stabilised yttria + neodymia ZrO_2 over yttria-stabilised ZrO_2, as discussed earlier, is the larger lattice parameter of double stabilised ZrO_2 as a result of the incorporation of Nd^{3+} ions and the absence of a thermally activated martensitic transformation above room temperature. This is confirmed by the peak shift of the t-ZrO_2 phase towards a lower 2θ angle (Figure 9a) whereas the m-ZrO_2 phase shows no peak shift (Figure 9b). At the same t-ZrO_2 phase transformability, the volume expansion resulting from the t-ZrO_2 phase transformation in doubled stabilised ZrO_2 composites will therefore be larger than in the yttria-stabilised equivalent implying a higher residual compressive stress on the propagating crack tip and concomitantly higher fracture toughness.

The slope of the LTD curve, which indicates the rate of t to m transformation or ZrO_2 degradation on the surface, is clearly higher for the ZrO_2-WC composites than for the ZrO_2-TiCN nanocomposites while the extension of degradation into the bulk material is slower in case of ZrO_2-WC. This is why, the (1Y, 1-2Nd) stabilised ZrO_2-WC composites remained intact even after 4000 minutes while (1Y,2Nd) ZrO_2-TiCN composites broke after 120 minutes. Most probably, the higher stiffness of ZrO_2-WC (~ 360 MPa) compared to ZrO_2-TiCN (~ 250 MPa) retards LTD progress. Matrix stiffness is known as one of the items that oppose ZrO_2 t to m transformation [9]. However, the (1Y-2Nd) stabilised ZrO_2 with a lower stiffness than the ZrO_2-TiCN composite remained intact after more than 1000 minutes of aging. More investigation is needed to fully understand the mechanism and the effect of secondary phases on the LTD.

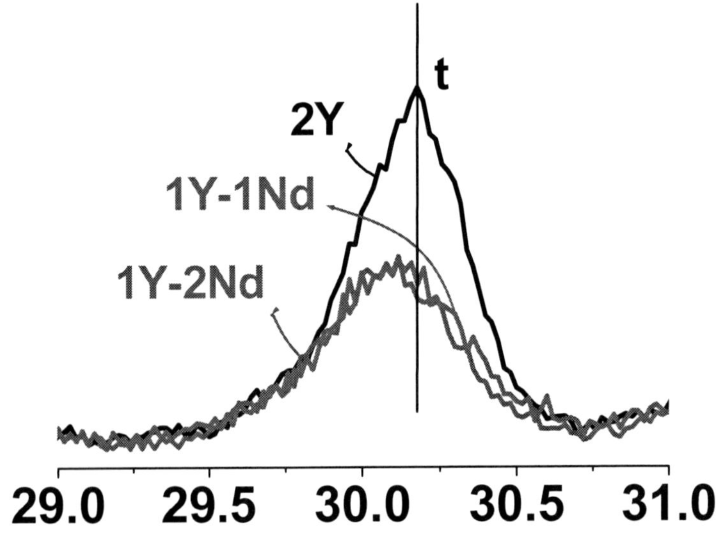

a)

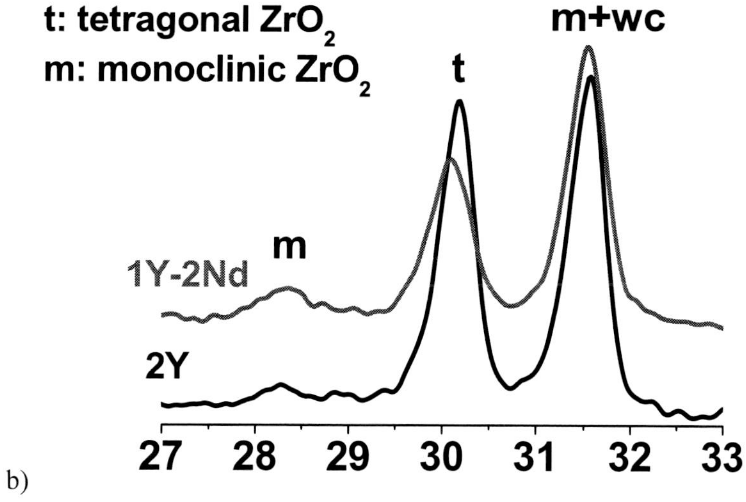

b)

Figure 9. XRD patterns of (1Y,1Nd)-, (1Y,2Nd)-, and 2Y-stabilised ZrO_2-WC (60/40) composites, PECS at 1450°C.

4. Conclusions

Despite the higher t-ZrO_2 phase transformability and fracture toughness, the resistance to hydrothermal degradation of 1 mol % Y_2O_3 + 1-4 mol % Nd_2O_3 co-stabilised ZrO_2 at 200°C was found to be better than for 2 and 3 mol % yttria-coated and 3 mol % yttria co-precipitated powder based ZrO_2. Although the surface of the (1Y,2-4Nd) co-stabilised ceramics partially transformed to m-ZrO_2, the material did not fracture after 1280 min at 200°C, whereas the yttria-stabilised grades completely degraded. The higher stability of the double stabiliser grades is attributed to the larger lattice parameter as a result of the incorporation of Nd^{3+} ions and the absence of a thermally activated martensitic transformation in the co-stabilised systems above room temperature.

The same concept explains the higher hydrothermal stability of double yttria + neodymia stabilised ZrO_2 based composites with 40 vol % TiCN and WC secondary phase compared to the yttria stabilised ones with the same amount of stabiliser content.

The hydrothermal stability of ZrO_2-NbC composites was found to increase with increasing NbC content, but could not be completely inhibited even when adding up to 70 vol % NbC. The higher hydrothermal stability with increasing NbC content was explained by the increased matrix stiffness, which constrains t to m ZrO_2 transformation.

References

[1] DJ Kim, HJ Jung, DH Cho, *Phase transformation of Y2O3 and Nb2O5 doped tetragonal zirconia during low temperature aging in air*, Solid State Ionics 80 (1995) 67-73.

[2] S Lawson, *Environmental degradation of zirconia ceramics*, J Eu Ceram Soc 15(1995) 485-502.

[3] J Chevalier, B Cales, J Drouin, *Low temperature aging of Y-TZP ceramics,* J Am Ceram Soc 82 (1999) 2150-2154.
[4] FF Lange, GL Dunlop, BI Davis, *Degradation during aging of transformation-toughened ZrO2-Y2O3 materials at 250°C,* J Am Ceram Soc 69 (1986) 237-240
[5] T Sato and M Shimada, *Transformation of yttria-doped tetragonal ZrO2 polycrystals by annealing in water,* J Am Ceram Soc 68 (1985) 356-356.
[6] X Guo, *Low temperature degradation mechanism of tetragonal zirconia ceramics in water: role of oxygen vacancies*, Solid State Ionics 112 (1998) 113-116.
[7] M Yoshimura, T Noma, K Kawabata, S Somiya, *Role of H2O on the degradation process of Y-TZP,* J Mat Sci Lett 6 (1987) 465-467.
[8] H Schubert and F Frey, *Stability of Y-TZP during hydrothermal treatment: neutron experiments and stability considerations*, J Eu Ceram Soc 25 (2005) 1597-1602.
[9] RHJ Hannink, PM Kelly and BC Muddle, *Transformation toughening in zirconia-containing ceramics,* J Am Ceram Soc 83 (2000) 461-487.
[10] T Xu, J Vleugels, O Van der Biest, Y Kan and P Wang, *Phase stability and mechanical properties of TZP with a low mixed Md2O3/Y2O3 stabiliser content,* J Eu Ceram Soc 26 (2006) 1205-1211.
[11] T Xu, J Vleugels, O Van der Biest, P Wang, *Fabrication and characterization of (Nd,Y)-TZP ceramics from stabiliser-coated nanopowder,* J Mater Lett 58 (2004) 3353-3357.
[12] Yuan, Z. X., Vleugels, J and Van der Biest, O., *preparation of Y2O3-coated ZrO2 powder by suspension drying.* J. Matter. Sci. Lett., 2000, 19, 359-361.
[13] L Jung-Dong, D Jens-Gong, L Chung-Lon, *Mechanical properties and resistance to hydrothermal aging of ceria and yttria-doped tetragonal zirconia ceramics,* Materials Chemistry and Physics 77 (2003) 808-818
[14] S Lawson, *Environmental degradation of Zirconia ceramics (review),* J Eur Ceram Soc 15 (1995) 485-502
[15] J Chevalier, L Gremillard and S Deville, *Low temperature degradation of Zirconia and implications for biomedical implants*, Ann Rev Mater Res 37 (2007) 1-32.
[16] M Yoshimura, T Noma, K Kawabate and S Somiya, *Role of H2O on the degradation process of Y,* J Mater Sci Lett 6 (1987) 465-7.
[17] S Salehi, O Van der Biest, J Vleugels, *Y2O3 and Nd2O3 co-stabilized ZrO2-WC composites* J Mater Sci 43 (2008) 5784-5789.

In: Handbook of Research on Chemoinformatics ...
Editor: A.K. Haghi
ISBN: 978-1-62100-998-6

Chapter 8

THE CHEMICAL TOPOLOGY OF CARBON ALLOTROPES AND THE EMERGENCE OF SPIRO QUANTUM CHEMISTRY

Michael J. Bucknum and Eduardo A. Castro*
(2) Research Institute of Theoretical and Applied Physical-Chemistry (INIFTA), Sucursal 4, Casilla de Correo 16, La Plata 1900, Buenos Aires, Argentina

PART I: SPIRO QUANTUM CHEMISTRY

ABSTRACT

This section has described the structure of a hypothetical 3-,4-connected net termed glitter. This is a model of an allotrope of carbon in the form of a synthetic metal. That paper pointed to the importance of through-space p_σ interactions of adjacent olefin units in the net in understanding the electronic structure at the Fermi level. The present communication elucidates the role of spiroconjugation in understanding features of the electronic band structure and density of states of glitter. With this analysis of spiroconjugation in the 1-dimensional polyspiroquinoid polymer and the 3-dimensional glitter lattice, the foundations have been laid for a new type of quantum chemistry herein called spiro quantum chemistry. Spiro quantum chemistry complements traditional quantum chemistry which is focused on linear polyenes, circular annulenes, polyhexes, 2-dimensional graphene sheets and related structures including fullerenes, by focusing on spiroconjugated hydrocarbon structures in 1- ,2- and 3-D, including linear spiro[n]quinoids and polyspiroquinoid in 1D, circular cyclospiro[n]quinoids, spiro[m,n]graphene fragments and spirographene in 2D and [m,n,o]glitter fragments and glitter in 3D.

*Corresponding author e-mail: eacast@gmail.com

Introduction

Recently the structure of the all-carbon "glitter" lattice [1] was described along with B_2C and CN_2 phases adopting this lattice. Glitter is a hypothetical tetragonal allotrope of carbon. The geometrical structure of the lattice (space group $P4_2/mmc$, # 131) is shown in Figure 1.

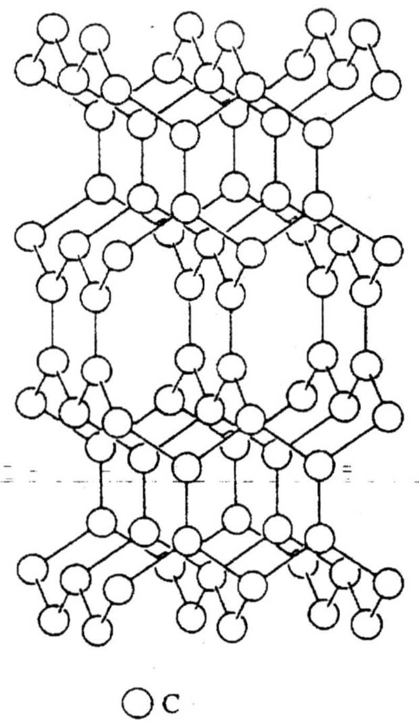

Figure 1. Structure of the glitter model in space group P42/mmc.

The dimensions of the lattice are **a** = 2.53 Å and **c** = 5.98 Å, these are constrained by the geometry of the 1,4-cyclohexadiene molecule upon which the structure is based. [2] It is a 3-,4-connected net [3] containing trigonal and tetrahedral atoms in a ratio of 2:1, with a calculated density of 3.12 g/cm^3; intermediate between graphite and diamond. Apparent in the structure are the ethylenic columns which run perpendicular to each other to generate the 3-dimensional lattice.

Merz et al. [4] have described the electronic band structures of several related 3-,4-connected nets each possessing stacked olefin units. Evolution of π-π^* band overlap with the separation between adjacent units in an olefin stack is presented in their paper, this diagram is shown in Figure 2. Note the touching of the π and the π^* bands at an interaction distance of about 2.5 Å. From this diagram it is clear that the approximate density of states profile of glitter will be that of a metal, with a small π-π^* band overlap at the Fermi level.

The structure shown in Figure 1 also contains the 1,4-cyclohexadieneoid molecular units linked through their tetrahedral vertices to adjacent rings. These linkages form chains. Each such chain is termed a "polyspiroquinoid" substructure of the "glitter" lattice. The electronic structure of a polyspiroquinoid model is described below in connection with the concept of spiroconjugation.

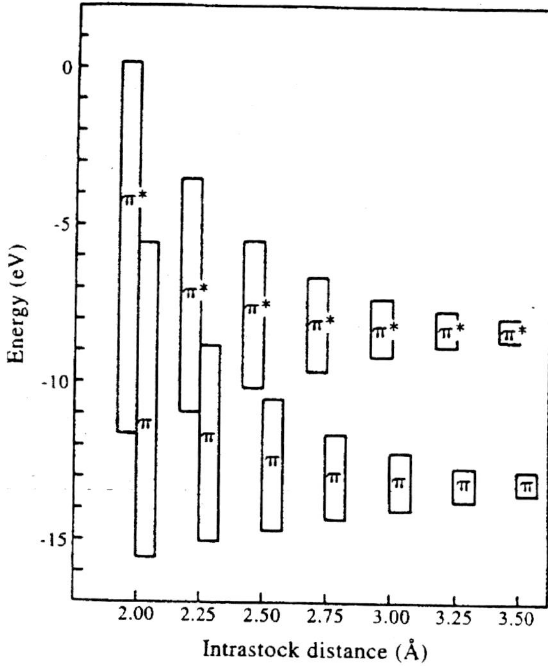

Figure 2. Evolution of π-π* overlap with the separation between adjacent units in an olefin stack.

Band structure calculations for this net were carried out using the extended Hueckel method adapted for application to extended structures. [5] Electronic band structures of various sublattice models, including the polyspiroquinoid one, were presented in the original report of the glitter structure [1] in order to clarify the importance of through-space interactions in these lattices.

1. Spiroconjugation in 1-Dimension and Polyspiroquinoid

Spiroconjugation [6] is claimed to occur in the electronic structure of the polyspiroquinoid chain. [1] This type of through-space interaction can be considered in terms of 4 p_{π} atomic orbitals held "spiro" to each other about a tetrahedral C atom (see Figure 3). These 4 orbitals interact to form 4 combinations of differeing symmetry with respect to the two perpendicular mirror planes dividing the tetrahedral C atom. One combination is symmetric under reflection in each mirror plane, labeled (SS), and there are also (SA), (AS) and (AA) combinations. The (AA) combination is shown, it is of the proper symmetry for a bonding spiro-type interaction.

From the original analyses of the effect of spiroconjugation in discrete molecular systems, including spiro[4.4]nonatetraene [7] the importance of it in stabilizing (and destabilizing) a frontier molecular orbital was of primary interest. This work was based upon electronic spectra, chemical reactivity and electronic structure calculations [6]. From this background, in the original paper on the glitter structure [1], only the lowest-lying π* band, the LUCO (lowest unoccupied crystal orbital) was considered in assessing the importance of through-space p_{spiro} interactions in the 1-dimensional polyspiroquinoid substructure of glitter.

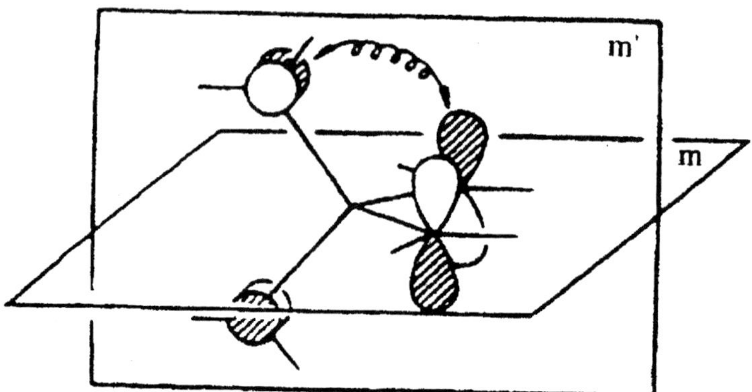

Figure 3. "Spiro" molecular orbitals derived from 4 pspiro combinations. The (AA) combination is shown.

The interaction diagram for spiro[4.4]nonatetraene, reported by Duerr and Kober [7h], is shown in Figure 4. The diagram reveals that in a situation involving 8 p_π orbitals (4 interacting π systems as opposed to 4 interacting p orbitals, as in Figure 3), there are generated 2 sets of 4 butadieneoid π systems. These 4 butadieneoid π systems are of (AS), (AA), (AS) and (AA) symmetry, unlike the archetypal system diagrammed in Figure 3. Interaction of these 4 π systems through the spiro C atom, yields 8 molecular orbitals over the skeleton of the spiro[4.4]nonatetraene molecule.

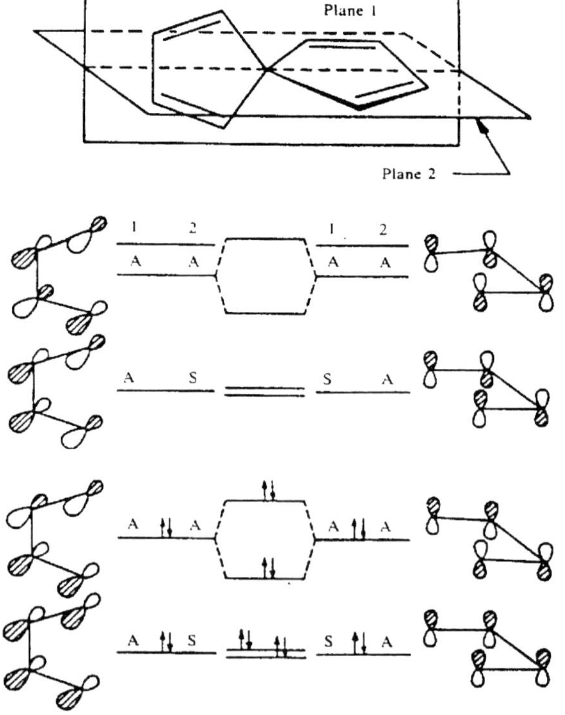

Figure 4. Molecular orbitals of spiro[4.4]nonatetraene.

Note the presence of the two (AA) combinations, one derived from the interaction of the HOMO (highest occupied molecular orbital) π levels of butadiene and one derived from the interaction of the LUMO+1 (next highest to the lowest unoccupied molecular orbital), π^* levels of butadiene. There are thus 4 levels in which the effects of spiroconjugation could be important, 2 down in the bonding manifold and the other pair in the anti-bonding manifold. The splitting of these pairs of (AA) combinations results from one spiro[4.4]nonatetraene orbital carrying a bonding p_{spiro} interaction and its sibling carrying an anti-bonding p_{spiro} interaction.

An analogy exists between the interaction diagram of the spiro[4.4]nonatetraene molecule and the unit cell of the polyspiroquinoid chain. Certain of the energy levels in both of these systems are comprised of 8 p_π atomic orbitals present in each structure. This analogy will be followed in the discussion below which describes the stabilizing effects of spiroconjugation in the electronic band structure of polyspiroquinoid.

Figure 5 shows the structure of the model polyspiroquinoid chain (H atoms added to the glitter fragment to reach a realistic structure) and Figure 6 is its electronic band structure. To the right in this diagram are appended the levels of the 3,3,6,6-tetramethyl-1,4-cyclohexadiene molecule. This molecule is a reasonable model for the polymer, but one lacking the spiroconjugation interaction. The 2 π and 2 π^* levels in the molecule have been indicated alongside the 4 π and 4 π^* bands of polyspiroquinoid.

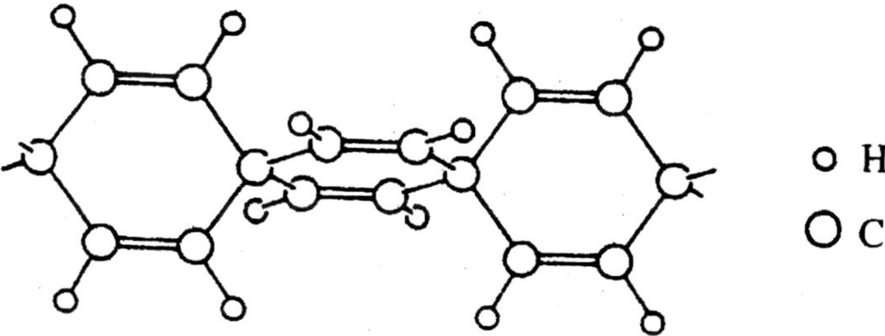

Figure 5. Structure of the polyspiroquinoid model.

As pointed out previously, [1] the lowest-lying π^* band in the electronic structure of the chain is of the proper symmetry, (AA), for there to be a bonding p_{spiro} interaction in that band. In addition, the lowest-lying π band is also of (AA) symmetry, so that there is a bonding p_{spiro} interaction in that occupied band as well. The relative importance of the effects of spiroconjugation in the chain will be posed with respect to the stabilization of the occupied π band compared to the energy level of the discrete π molecular orbital of 3,3,6,6-tetramethyl-1,4-cyclohexadiene from which it is derived.

In analogy with spiro[4.4]nonatetraene, the 8 p_π atomic orbitals in the unit cell of polyspiroquinoid generate 4 π crystal orbitals and 4 π^* crystal orbitals in its electronic band structure. At the zone center, half of these 8 crystal orbitals are constructed from real coefficients of the same magnitude at each p_π atomic site in the unit cell. Because of their importance in the analysis of spiroconjugation in polyspiroquinoid, these 4 (AA) crystal orbitals are sketched exactly as they occur at the zone center, in Figure 7. These are analogous to the 4 (AA) combinations from the spiro[4.4]nonatetraene analysis. They occur as band #18 (π_1), band #22 (π_2), band #25 (π_5^*) and band #28 (π_8^*).

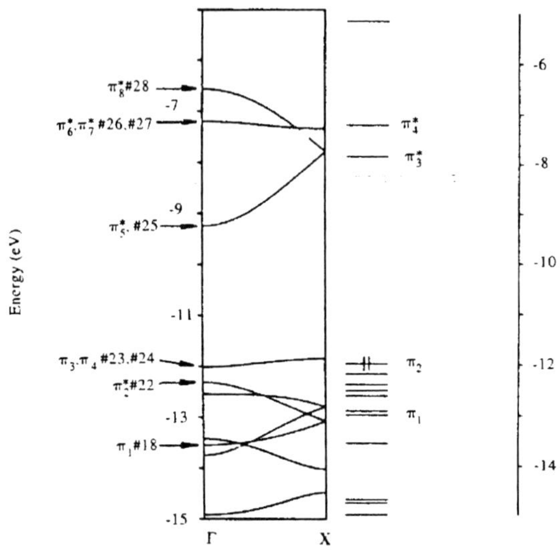

Figure 6. Electronic band structure of polyspiroquinoid and molecular orbitals of 3,3,6,6-tetramethyl-1,4-cyclohexadiene with π and π* levels indicated.

The 4 (AS) combinations occur in degenerate band pairs (compare to the orbital interaction diagram of spiro [4.4]nonatetraene) across the Brillouin zone of the chain. These are constructed from real coefficients of the same magnitude and sign at each p_π atomic site in one or the other ring of the polyspiroquinoid unit cell. Every other ring in the polymer possesses these π and π* type interactions. The π and π* interactions in half the unit cell are involved in σ or σ* interactions with the adjacent ring, which maintain the (AS) symmetry of the crystal orbital. At the zone center they occur as bands #23 (π_3) and #24 (π_4), (the HOCO, highest occupied crystal orbital, levels) and as bands #26 (π_6*) and #27 (π_7*), (the LUCO+1 levels).

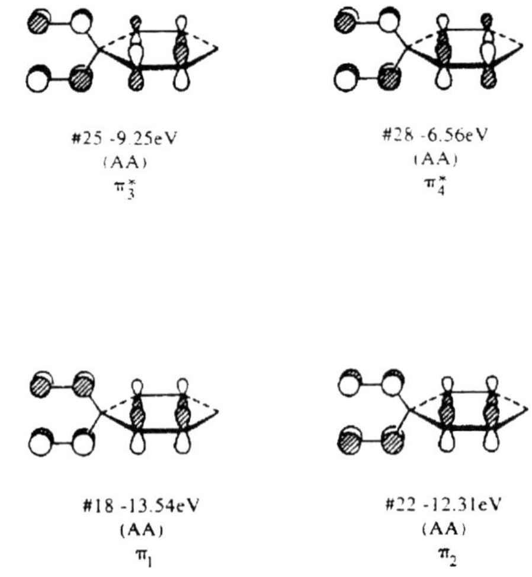

Figure 7. The pπ crystal orbitals of polyspiroquinoid sketched at the center of the Brillouin zone.

One can analyze the attendant orbital interactions of the (AA) combinations graphically from the sketches. In crystal orbital #18, the 4 bonding p_π combinations in the unit cell also possess a bonding p_{spiro} interaction about each of the spiro carbons in the chain. Comparison with crystal orbital #22, the sibling π band to crystal orbital #18, reveals their only difference is the presence of an anti-bonding p_{spiro} interaction in the higher-lying π band. Their energy difference at the zone center is 1.23 eV, this is the actual magnitude of the splitting of the (AA) combinations; this was predicted qualitatively in the orbital interaction diagram shown in Figure 4.

In the 3,3,6,6-tetramethyl-1,4-cyclohexadiene model molecule, the lower-lying π (π_1) level is the "out-of-phase" π combination, it is of (AA) symmetry with respect to the 2 mirror planes dividing the spiro carbon atoms in the ring. Its sibling level (π_2) is the "in-phase" π combination, it is of (AS) symmetry. The 2 π (AA) levels in polyspiroquinoid are derived from the pure "out-of-phase" π combination of this model molecule, and the 2 (AS) levels are derived from the pure "in-phase" π combination. Crystal orbital #18 (π_1) occurs at –13.54 eV at the zone center and crystal orbital #22 (π_2) occurs at –12.31 eV. Therefore the spiroconjugation band is stabilized by 0.55 eV with respect to the lower-lying π level in 3,3,6,6-tetramethyl-1,4-cyclohexadiene and its sibling π band, with anti-bonding p_{spiro} interactions, is destabilized by 0.68 eV.

Tracing the evolution of these bands along the symmetry line of polyspiroquinoid, one sees the 2 sibling bands meet at the zone edge where they are degenerate (see Figure 6). This degeneracy occurs because of the presence of a 4_2 screw axis in the polyspiroquinoid unit cell. The band structure is "folded" in half due to the 4_2 screw axis. Across the Brillouin zone crystal orbital π_1, the spiroconjugation band, is stabilized with respect to the lower-lying π level in 3,3,6,6-tetramethyl-1,4-cyclohexadiene (π_1). At the zone edge, the degeneracy at –13.07 eV is just 0.08 eV below the corresponding molecular orbital (π_1) in the 3,3,6,6-tetramethyl-1,4-cyclohexadiene model molecule.

Figure 8 shows the density of states (DOS) diagram along with the band structure diagram (Figure 6) of polyspiroquinoid.

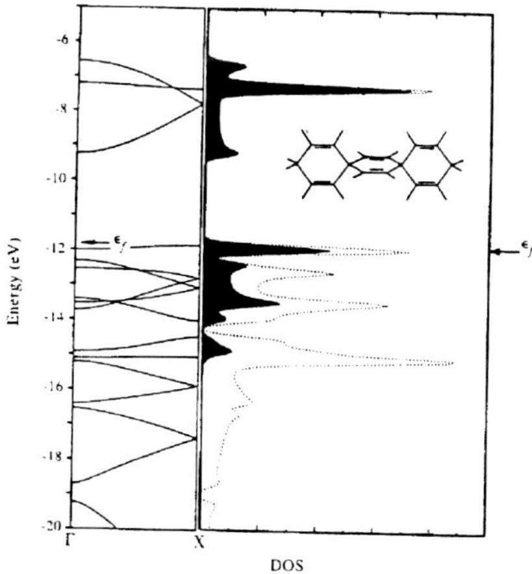

Figure 8. Electronic band structure and density of states (DOS) of the polyspiroquinoid model; shaded area indicates contribution of pπ orbitals to the total DOS.

The shaded area is the contribution of p_π orbitals to the total DOS. It has a "spike" at an energy which corresponds to the spiroconjugation band at the zone center (-13.5 eV). This spike indicates there is a large density of states in the electronic structure of the polymer at that energy. Evidently from the DOS diagram, there is not a corresponding spike coincident with the energy of the sibling π band which possesses anti-bonding p_{spiro} interactions at the zone center. This is consistent with the small degree of dispersion (flatness) of band #18 in the neighborhood of the zone edge.

2. Spiroconjugation in 3-Dimensions and Glitter

In the full 3-dimensional glitter lattice shown in Figure 1, it is obvious the polyspiroquinoid chains are linked together along the [100] and [010] directions of the lattice. Although through-space bonding p_σ interactions are primary in the analysis of what occurs at the Fermi level of glitter, it turns out that the spiroconjugation effects seen in the 1-dimensional polyspiroquinoid substructure will carry over into the 3-dimensional structure.

As is implied in the model structural drawing on the inset of Figure 9, through-space bonding p_σ interactions are responsible for reducing the band gap (see Figure 2) in the 1-dimensional polycyclophane substructure, which possesses the p_σ interactions uniquely among the 1-dimensional substructures.

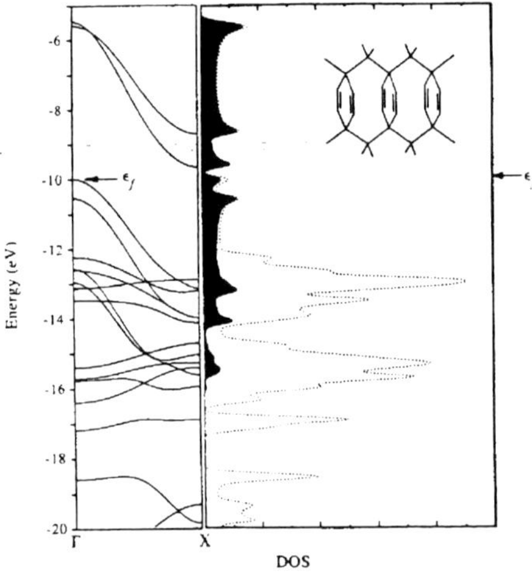

Figure 9. Electronic band structure and density of states (DOS) of the polycyclophane structure. Shaded area indicates the contribution of pπ orbitals to the total DOS. Structural model shown on onset.

Clearly in this substructure, which from its geometry is absent orbital interactions connected with spiroconjugation, the π and π* crystal orbitals come in pairs, the HOCO-1 and the HOCO as the π bands, and the LUCO and the LUCO+1 as the π* bands. These bands run down from the zone center, where they occur in nearly degenerate pairs, out to the zone edge, as shown in Figure 9.

Figure 10 shows the band structure of glitter, as there are 6 C atoms in the unit cell it consists of 24 bands, 12 of these bands are fully occupied. In order to assess the importance of through-space interactions in the lattice, including the importance of spiroconjugation, it is only necessary to consider the crystal orbitals derived from combinations of the p_π atomic orbitals present in the unit cell of glitter. These 4 p_π atomic orbitals, one such atomic orbital for each of the 4 trigonal planar C atoms in the unit cell, combine together to form 2 π crystal orbitals and 2 π^* crystal orbitals in the electronic band structure of the glitter lattice.

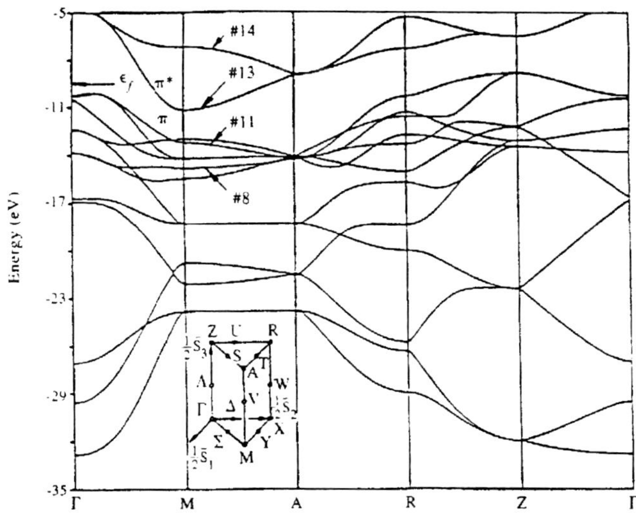

Figure 10. Electronic band structure of glitter. The π and π^* bands are indicated at symmetry point M of the Brillouin zone.

In analogy with the polycyclophane substructure, it would be expected that these π and π^* crystal orbitals would bracket the Fermi energy; that is they would occur as the HOCO-1, HOCO, LUCO and LUCO+1 (see Figure 10). In the original report of the electronic structure of glitter, the authors pointed to the importance of through-space p_σ interactions, occurring together with through-bond p_π interactions, at the short interaction distance of 2.53 Å present in the unit cell.

Comparison of the Brillouin zone of glitter to the unit cell of glitter in direct space, shown on the inset of Figure 10, reveals that symmetry point M is analogous to the zone center in the Brillouin zone of the polyspiroquinoid model. The symmetry line from M to A corresponds to the symmetry line followed from the zone center to the zone edge in the electronic structure of the polyspiroquinoid model. The π and π^* bands in glitter, the π bands labeled as #8 and #11 and the π^* bands labeled as #13 and #14, at symmetry point M in the diagram, become degenerate pairs at symmetry point A. These degeneracies occur because of the presence of a 4_2 screw axis in the unit cell of glitter.

At symmetry point M in the Brillouin zone of glitter these 4 crystal orbitals enter with real valued coefficients of the same magnitude. It is possible to carry out a graphical examination of these 4 crystal orbitals derived from the 4 p_π atomic orbitals of glitter. They may be sketched exactly as they appear in the Brillouin zone at symmetry point M. From these sketches, which are shown in Figure 11, the attendant orbital interactions may be analyzed graphically.

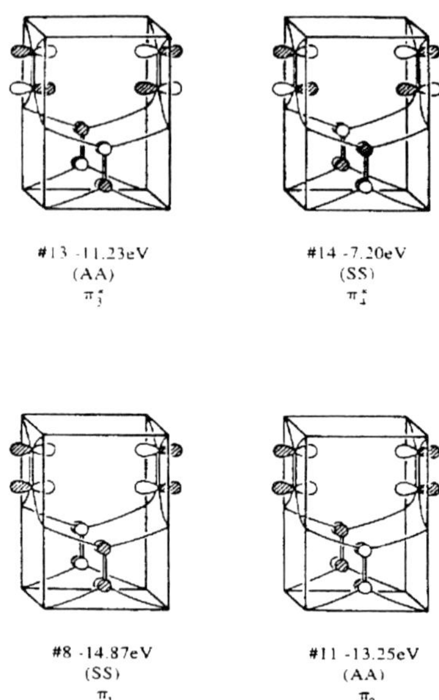

Figure 11. 4 pπ crystal orbitals of glitter at symmetry point M in the Brillouin zone.

Labeling the 4 π crystal orbitals at high symmetry point M in the Brillouin zone, using their symmetry with respect to the 4_2 screw axis and the **c** glide plane of the unit cell, there are 4 combinations: π_1 (SS), π_2 (AA), π_3* (AA) and π_4* (SS). Using the 2 mirror planes along [001] to assign the symmetries of the π crystal orbitals, instead of their behavior with respect to the space symmetry elements, they would all be of symmetry (SS), this represents an inversion of symmetry from the polyspiroquinoid (AA) combinations. Had the unit cell been chosen with [100] and [010] rotated through 45° from the present unit cell, all 4 combinations would be of (AA) symmetry with respect to the 2 mirror planes along [001]. This is entirely analogous to the 4 (AA) combinations of the polyspiroquinoid model, although not strictly homologous to the electronic structure of the polymer, as is discussed later. At any event, the present choice of unit cell makes visualization of the orbital interactions more clear to the reader.

From these sketches, one can see the presence of p_π interactions within the unit cell and also the presence of inter-cell and intra-cell p_σ and p_{spiro} interactions. Band #8 is the lowest-lying π crystal orbital (-14.87 eV), labeled π_1 (SS). It possesses through-space bonding p_{spiro} interactions and lies 1.67 eV below the next highest π crystal orbital, band #11 labeled π_2 (AA) (-13.25 eV) at symmetry point M. This splitting is analogous to the splitting of the bonding (AA) combinations in polyspiroquinoid. Bands #13 and #14 are the 2 π* crystal orbitals at M, they differ only in that the lower-lying of the pair has bonding p_{spiro} interactions, they have an energy separation of over 4 eV.

At this symmetry point in the Brillouin zone, the spiroconjugation effects present in the unit cell of glitter provide a stabilization energy of 1.93 eV compared to the energy of the "out-of-phase" π combination (π_1) of 3,3,6,6-tetramethyl-1,4-cyclohexadiene (see Figure 6). The destabilized sibling π band (AA), at –13.25 eV, is itself 0.26 eV stabilized with respect to the "out-of-phase" π combination of the 3,3,6,6-tetramethyl-1,4-cyclohexadiene model molecule. Evidently, the p_σ interactions are quite important in stabilizing this π_2 (AA) crystal orbital as well.

An indirect band crossing of the lower-lying π^* band, the LUCO, with the HOCO and 2 other bands, is present in the band structure of glitter (see Figure 10). Of the 2 π^* crystal orbitals present in the electronic structure of glitter, the π^* crystal orbital which is the LUCO has through-space bonding p_{spiro} interactions (along with the p_σ interactions) which cause this band to dip to a minimum that crosses indirectly the lower-lying occupied bands in glitter. This is pictured in Figure 11 and thus explains the origin of the synthetic metallic status adopted by this hypothetical allotrope of C.

Figure 12 shows the electronic band structure for the glitter lattice in the energy window from –15 to –5 eV. Inspection of this calculated band diagram shows that the lowest-lying π band, band #8, occurs at its lowest energy point at symmetry point M in the Brillouin zone, as does the lower-lying π^* crystal orbital. The lowest-lying π band is the HOCO-4, not the expected HOCO-1.

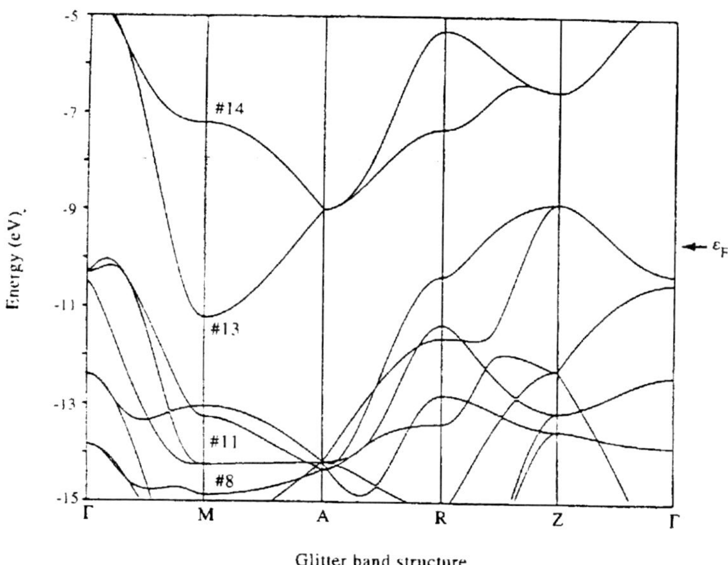

Figure 12. Electronic band structure of glitter in the energy window from –15 to –5 eV.

It occurs as such through the entire Brillouin zone of glitter. Band #11, refer to the corresponding sketch in Figure 10, is the next highest π band, it forms a degenerate pair as the highest occupied crystal orbital along the symmetry line from Z to Γ in the Brillouin zone of glitter. Inspection of the Brillouin zone (Figure 12) reveals that the 2 π crystal orbitals maintain an energy separation of about 1 eV throughout most of the reciprocal space of glitter

Note also in Figure 12, the small dispersion (flatness) of band #8 about symmetry point M in the Brillouin zone, this gives rise to the secondary peak in the DOS of glitter (see Figure 13), at an energy of about –15.0 eV. The shaded area indicates the contribution of p_π orbitals to the total DOS. The primary peak occurring at about –13.0 eV nearly coincides with the flatness of the sibling π crystal orbital, band #11, about this symmetry point in the Brillouin zone of glitter.

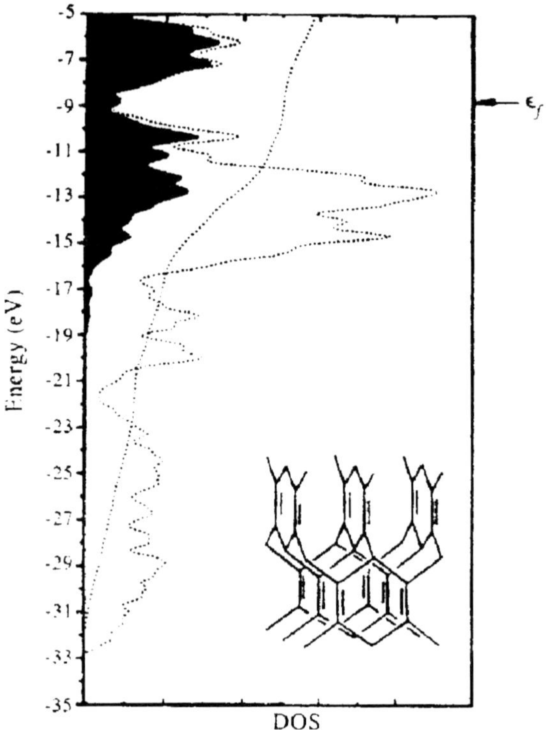

Figure 13. Density of states (DOS) of glitter. Shaded area indicates the contribution of pπ orbitals to the total density of states.

3. The Special Crystal Orbital of Glitter

Inspection of the atomic orbital composition of band #8 of glitter, the lowest-lying π crystal orbital, at symmetry point M in the Brillouin zone, reveals it has all 4 p_π coefficients entering with the *same magnitude and sign.* Because of the presence of a bonding p_π interaction, 2 bonding p_σ interactions and 4 bonding p_{spiro} interactions about *every* p_π atomic orbital in the unit cell of glitter, this is termed its special crystal orbital. This is diagrammed in Figure 14. The effects of spiroconjugation along with the other interactions present in this 3-,4-connected network may confer it with an unusual electronic stability akin to the aromaticity observed in the planar polyhexes and 2D graphene sheet of ordinary quantum chemistry.[8]

The special crystal orbital can be viewed in sections through several unit cells perpendicular to the (001) plane of the tetragonal structure, and with the spiro atoms flattened out for clarity, as is shown in Figure 15 (this diagram is applicable to the view of the lower-lying π* crystal orbital, the LUCO, perpendicular to (001) as well, however alternating sheets would be out of phase along the 3rd dimension in the LUCO). Note the resemblance of this pattern to that of a standard checkerboard square, it arises as a consequence of the tetragonal symmetry of the glitter structure and from the nature of the p orbital interactions admitted by the glitter lattice. The constructive interactions present in the unit cell involve intra-cell interactions: p_π, p_σ and p_{spiro}; and inter-cell interactions of the same nature

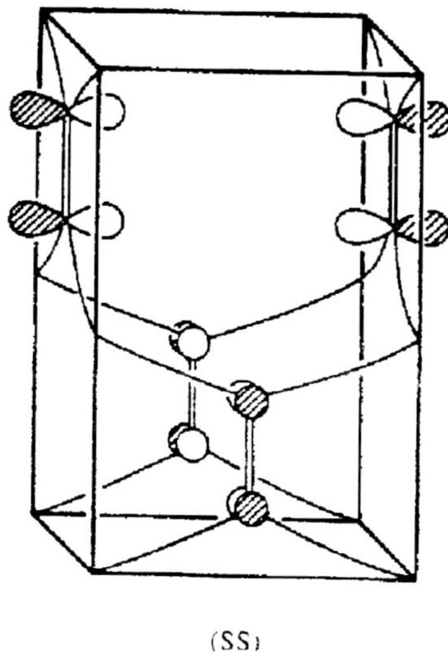

Figure 14. The special crystal orbital of glitter.

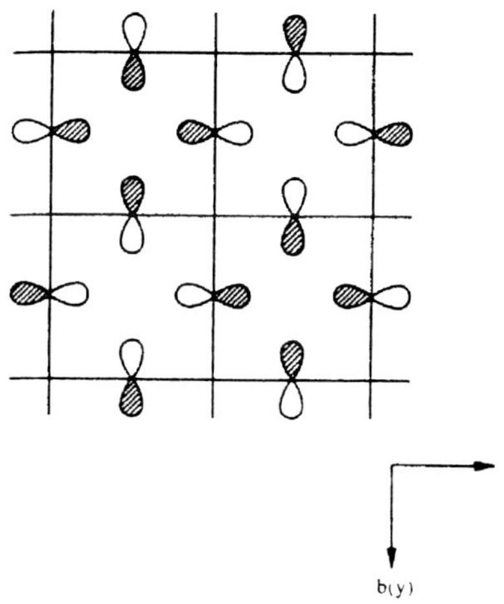

Figure 15. "Sections" of the special crystal orbital along [100] and [010] lattice directions. View is perpendicular to (001).

4. Exo- and Endo-Spiroconjugation and Diffraction

Self assembly involving 4 allene units per reaction, in the presence of a catalytic amount of the organometallic reagent bis-(triphenylphosphite)-nickeldicarbonyl, has been reported to produce 1,3,5,7-tetramethylenecyclooctane. [10] Synthesis of 1,3,5,7-tetramethylenecyclooctane represents the 1st example of a molecular system with the possibility of exhibiting the effect of endo-spiroconjugation. Well known in structural chemistry are the homologies provided by the hydrocarbon molecules naphthalene and adamantane, which have very nearly the carbon-carbon bond lengths and bond angles, respectively, of their homologous extended structures: graphite and diamond. [9] From this perspective, the synthesis of 1,3,5,7-tetramethylenecyclooctane provides the opportunity to have an homologous hydrocarbon fragment of the structure of glitter which can be studied experimentally.

Endo-spiroconjugation, potentially present in the 1,3,5,7-tetramethylenecyclooctane molecule, is precisely the mechanism of spiroconjugation that would occur intracellularly and extracellularly in the glitter lattice, please refer to the previous sections for details. Literature previously cited on aspects of spiroconjugation [7] concerned the related effect of exo-spiroconjugation, the interaction diagram of which is shown in Figure 3 (interaction of 4 p orbitals) and Figure 4 (interaction of 4 π orbitals). As discussed in the 2nd section, in the hypothetical polyspiroquinoid polymer, exo-spiroconjugation is operative.

Ideally, endo-spiroconjugation involves 4 p orbitals that point along lines, through space, which represent a pair of *edges* of a perfect tetrahedron, held at right angles to each other. Oppositely, exo-spiroconjugation, as pictured in Figure 3, involves 4 p orbitals which are centered at each *vertex* of a tetrahedron and point at right angles, in 2 different pairs, respectively, each to one of a corresponding pair of orthogonal tetrahedral edges. Thus, the tetrahedron in exo-spiroconjugation is formed from 4 sp^3 hybrid orbitals extending from the spiro C atom to each of the atoms possessing the interacting p orbital. Quite in contrast, the p orbitals at a site of endo-spiroconjugation are cenetered on a point in space. From this subtle difference it is apparent that, site-by-site, the p orbital endo-spiroconjugated sees each of the 2 p_{spiro} interactions of its counterpart p orbital that is exo-spiroconjugated; *and simultaneously it sees a* p_σ *interaction.*

In the structure of 1,3,5,7-tetramethylenecyclooctane, ideally the effect of endo-spiroconjugation should manifest itself in the geometry of the skeletal cyclooctane ring system; the 4 trigonal atoms in the ring should inscribe a perfect tetrahedron inside the octagonal cradle. At the center of the inscribed tetrahedron, a significant π electron density should be present from the 4-center interaction of endo-spiroconjugation. As the scattering power for x-rays depends upon the electron density, diffraction patterns obtained from molecular crystals of the 1,3,5,7-tetramethylenecyclooctane molecules may indicate weak reflections corresponding to the electron density $\rho_{spiro}(x,y,z)$ at the fractional coordinates of the unit cell specifying the center of the inscribed tetrahedron in the cyclooctane cradle (the center of symmetry of the molecule). Alternatively, electron density maps of the structure of 1,3,5,7-tetramethylenecyclooctane calculated by direct methods from the original diffraction data would reveal, quantitatively, the magnitude of π electron density at the fractional corrdinates corresponding to the center of symmetry of the 1,3,5,7-tetramethylenecyclooctane molecule.

Reflections due to π electron density concentrated at the center of symmetry of the 1,3,5,7-tetramethylenecyclooctane molecule would be weak for two principal reasons; the scattering power of the skeletal C atoms would be approximately 6 at low values of Θ; 6 being the number of electrons in C, whereas $\rho_{spiro}(x,y,z)$ could be at most approximately 2 at low values of Θ. The value 2 at low Θ is thus identified with the electronically unstable case that the π electron pair is cradled at the center of the inscribed tetrahedron. Intensity from $\rho_{spiro}(x,y,z)$ would be diminished substantially as the inscribed tetrahedron distorted from local T_d symmetry. As the non-ring C's in 1,3,5,7-tetramethylenecyclooctane are not stabilized by the benefit of endo-spiroconjugation, it is to be expected that the molecule will distort; it will maintain its principal S_4 axis, although it will distort towards a flattening out of the cyclooctane skeleton. Consequently, the local T_d symmetry of the inscribed tetrahedron will be lowered to one of D_{2d} symmetry and the scattering power at the endo-spiro center, $\rho_{spiro}(x,y,z)$, will be greatly diminished.

Construction of the model of the glitter lattice appealed to the geometry of the 1,4-cyclohexadiene molecule, information provided from diffraction studies [2]. The carbon-carbon double bonds of the molecule are not conjugated and this appears in the bond alternation, the carbon-carbon single bonds being 1.51 Å and the carbon-carbon double bonds being 1.35 Å (in naphthalene the carbon-carbon bond lengths are approximately 1.40 Å, and in adamantane the carbon-carbon bond lengths are approximately 1.54 Å, respectively). Similarly, the bond angles are unsymmetrical in 1,4-cyclohexadiene with the C=C-C angles being 123° (ideal trigonal angle is 120°) and the adumbrated C-C-C angles are 114° (ideal tetrahedral angle is 109°28′).

Bond length and bond angle alternation present in the 1,4-cyclohexadiene molecule, and consequently the glitter lattice which is based upon it, distinguishes glitter from the metrically ideally regular sibling structures: graphite and diamond. See a later section for a further description of this metrical difference. Also discussed later, the standard topological indexes, polygonality and connectivity, distinguish glitter further as a topologically irregular (i.e. Wellsean) structure, while two of its siblings, graphite and diamond, are regular 2-dimensional and 3-dimensional structures, respectively; and the 3rd group of its siblings, the crystalline fullerenes, are semi-regular polyhedra (i.e. Archimedean).

Apparently quite by accident, the irregular metrical properties of the glitter lattice; the bond angle and bond length alternation, concomitant with its "low" topology (see a later section), result in the occurrence of nearly perfect tetrahedra, centered in space, which are about 2.5 Å on an edge. Such tetrahedra, being centered in space as opposed to being centered at an ideally tetrahedral C atom, as in diamond for example, provide the vehicle within which the 4 center p orbital interaction of endo-spiroconjugation may occur. These tetrahedra possess nearly the full T_d symmetry that is evidently required for the optimum energetic stabilization associated with endo-spiroconjugation, refer to the earlier sections for a quantitative measure of this interaction energy.

Superimposed upon the unit cell of glitter; interpenetrating with it, which itself is outlined by the positions occupied by the 6 C atoms in $P4_2/mmc$ in glitter, are the set of points coincident with the spatial centers of the nearly perfect tetrahedra. These are the sites at which, through the effects of endo-spiroconjugation, are concentrated the π electron density from the 4 center, through-space interaction. The latter array constitutes a primitive tetragonal lattice of points (P4/mmm) which is half the volume of the unit cell of glitter, the secondary lattice having its origin centered at (1/2,1/2,0) of the primary glitter lattice. Each of these sites may have a scattering power approaching 2 at low x-ray scattering angles, Θ. In light of the T_d symmetry of these sites, and the more closely-spaced, regular array of such sites in the glitter lattice, compared to the situation in a molecular crystal of 1,3,5,7-tetramethylenecyclooctane discussed previously, it is to be expected that any weak reflections associated with x-ray scattering from the sites of $\rho_{spiro}(x,y,z)$ in glitter would be more evident than in a comparable diffraction pattern of a molecular crystal of 1,3,5,7-tetramethylenecyclooctane.

5. Exo- and Endo-Spiroconjugation and Chemistry

Williams and Benson [10] report aspects of the chemistry of 1,3,5,7-tetramethylenecyclooctane which provides some evidence for the proximity of π electrons within the cradle of the cyclooctane ring system. Effects on chemical reactivity due to exo-spiroconjugation have been referred to earlier [7]. It was hoped that evidence for the proximal interaction of adjacent, exocyclic methylene groups of the molecule would be evident through the formation of a transannular bond, in analogy with the cycloaddition reaction of tetracyanoethylene with bicycloheptadiene producing nortricyclene [11]. Cycloaddition of tetracyanoethylene with 1,3,5,7-tetramethylenecyclooctane falls within the class of symmetry-allowed, thermal s,s,s reactions; it is a $[_{\pi}2_s + _{\pi}2_s + _{\pi}2_s]$ cycloaddition reaction. [12]

Tetracyanoethylene was chosen as the reagent to establish the interaction of pairs of opposite double bonds, the ring carbon p orbitals of which point towards each other along a line corresponding to one of the two orthogonal edges of the tetrahedron inscribed in the cradle of the cyclooctane skeleton, described above. An approximately equimolar reaction mixture of the two was prepared in tetrahydrofuran (THF) at room temperature. An exothermic reaction occurred within minutes, and after an hour the product was isolated, crystallized from acetonitrile and from spectroscopis analysis was determined to have a 3,3,4,4-tetracyano-8,11-dimethylenetricyclo[4.3.3.0]-dodecane structure. As the author's pointed out,"this appears to be the first example in which transannular bond formation can be directly attributed to the *proximity* of exocyclic methylene groups" [10].

6. Distortions of the Glitter Lattice

Based upon the crystal structures of diamond and graphite [9], in which the bond lengths and bond angles are uniform within each structure, respectively, a distortion was performed on the glitter lattice such that the bond lengths were all fixed at 1.46 Å (instead of alternating between 1.35 Å and 1.51 Å) and the bond angles were all fixed at 120° (except around the circumference of the tetrahedral C atoms, which is constrained to about 105°). The distortion to a more metrically regular structure destabilized the lattice by 0.80 eV/C atom relative to the metrically irregular structure.

Evidently, as the metrical properties of the lattice are modified to be more uniform, the tetrahedral sites that are centered in space, associated with the 4 center p orbital interaction of endo-spiroconjugation, concomitantly undergo a lowering of their local symmetry from T_d to D_{2d}. This is consistent with the model of endo-spiroconjugation presented here, in which the ideality of the tetrahedron yields optimum stabilization from the effect of endo-spiroconjugation.

7. Synthesis of Glitter

Glitter is a 3-,4-connected net, so it is to be expected that a synthesis would be possible at conditions of temperature (T) and pressure (P) dictated by the phase boundary between graphite and diamond. However, the recent literature on C reflects the importance of kinetically driven syntheses, especially at conditions of T and P other than those along the graphite-diamond phase boundary. These include syntheses with vaporous carbon and hydrocarbon precursors in deposition apparatus [13]. A thermodynamic synthesis, in the vein of the original synthesis of diamond from graphite, could possibly be achieved in the presence of a suitable catalyst, such as a metallic solvent. Conventional high T-high P apparatus could be used for such a synthesis [14].

Identification of phases produced in such synthetic programs would principally include the use of diffractometry and spectroscopy. From the details of the electronic band structure of glitter reported herein, it should be possible to assign π-π^* electronic transitions through the analysis of the uv-visible absorption spectra, photelectron spectroscopy should elucidate features of the highest occupied crystal orbital. [15] With regard to diffraction evidence, theoretical diffraction patterns could be calculated for various versions of the crystal structure of glitter and these theoretical diffraction lines could be searched for in a synthesis experiment such as one employing a diamond anvil synthesis cell (DAC) and synchrotron radiation.

8. Topological Indexes of Carbon Allotropes and Glitter

Wells' seminal work on the structures of 2- and 3-dimensional nets and polyhedra was organized through the use of topological labels called Schlaefli symbols. [3] In such a scheme, there are two indexes, the polygonality represented by n and the connectivity represented by p. Each polyhedron, plane tessellation and 3-dimensional net has a Schlaefli symbol (n, p) to label its location in the space of topological structures in which Wells worked. [16] Interestingly, the Schlaefli symbols are rigorously determined for the convex polyhedra through a relation due to Euler.

Euler's work on the convex polyhedra resulted in an equation marking the origin of the discipline of topology, shown in one form as Equation 1 below. [17]

$$V - E + F = 2(\qquad 1)$$

Here V, E and F are the number of vertices, edges and faces, respectively, in the convex polyhedron. [18] Geometrical arguments can be used to transform Equation 1 into forms relating V, E and F; the primary topological indexes, to n, the polygonality and p, the connectivity; the latter being secondary topological indexes.

Each edge in a convex polyhedron is shared by 2 faces, therefore nF is the same as 2E. Each edge has 2 vertices, therefore pV is the same as 2E. By substitution, Euler's equation now reads:

$$^{2E}/_{n} - E + {}^{2E}/_{p} = 2 \qquad (2)$$

Rearrangement of Equation 2 into Equation 3 shows that the values n, p and E must be positive integers for the expression to have validity for polyhedra: [15]

$$^{1}/_{n} - {}^{1}/_{2} + {}^{1}/_{p} = {}^{1}/_{E} \qquad (3)$$

Further restrictions are imposed on the values of n and p in order to determine the solutions to Equation 3. In order for the number of edges, E, to be positive the values of n and p must be less than 6, and in fact n and p must be greater than 2 because of the impossibility of faces of zero area or spikes in the convex polyhedron.

Substituting each of the 9 combinations of 3, 4 and 5 into Equation 3, there are only 5 finite, rational solutions. The 5 are well known and are shown as the tetrahedron, t, the octahedron, o, the icosahedron, i, the cube, c, and the dodecahedron, d, in the table below:

Table 1. Schlaefli Symbols for Regular Structures

	p						
n	**3**	**4**	**5**	**6**	**7**	**8**	...
3	t	o	i	(3,6)	(3,7)	(3,8)	
4	c	(4,4)	(4,5)	(4,6)	(4,7)	(4,8)	
5	d	(5,4)	(5,5)	(5,6)	(5,7)	(5,8)	
6	(6,3)	(6,4)	(6,5)	(6,6)	(6,7)	(6,8)	
7	(7,3)	(7,4)	(7,5)	(7,6)	(7,7)	(7,8)	
8	(8,3)	(8,4)	(8,5)	(8,6)	(8,7)	(8,8)	

Of the other 4 non-solutions, from the square net (4, 4) substitution into Equation 3 indicates the number of edges, E, in this structure is infinite. Evidently, in (4, 4) the Euler equation reveals a transition to a 2-dimensional extended structure. [19] In (4, 5), (5, 4) and (5, 5) the non-solutions have values of –5, -5, and –10, respectively. Another transition has occurred, to a still higher dimensionality, specifically to 3-dimensional structures. (5, 5) corresponds to a 3-dimensional network that Wells first discovered in the 1960's. [3] For a sensible result, in which the number of edges, E, is positive, it is clear that a modification of the classical Euler equation, one applicable for 3-dimensional networks, is necessary. [20]

The structure of the graphite net is well known, its Schlaefli symbol is (6, 3), the polygonality is 6 and the connectivity is 3. Its place in the topological subspace comprised of regular structures, regular meaning the Schlaefli symbols n and p are integers, is to the right of the convex polyhedron (5, 3), which is the pentagonal dodecahedron, and to the left of (7, 3). Interestingly, (7, 3) represents a group of at least 4 unique, regular nets which Wells discovered. [3] To date, none of these nets in the family (7, 3) has yet been discovered to correspond to an actual crystal structure.

Regular structures are adjoined to their left or right, or adjacent to them vertically, by the semi-regular structures, semi-regular meaning that one of the Schlaefli indexes n and p is fractional and the other is an integer. One such semi-regular with contemporary importance is the Archimedean polyhedron called the truncated icosahedron. The Schlaefli symbol for this mathematical object, and for the Buckminsterfullerene molecule that is patterned after it, is given by ($5^{2/3}$, 3). It is intermediate in topology between the pentagonal dodecahedron (5, 3) and the graphite net (6, 3). Buckminsterfullerene is unique among the allotropes of C in that it is semi-regular.

Classical structures include the regular and Archimedean polyhedra; Archimedean meaning only n, the polygonality, is fractional. In the 19th century, Catalan identified the reciprocal polyhedra to the Archimedean polyhedra where n and p were exchanged with each other. Catalan polyhedra can be constructed from their reciprocals by joining the midpoints of each face of the Archimedean polyhedron to each other [21]. This reciprocal nature occurs also in 2 dimensions, one can easily visualize the reciprocals of (4, 4), (3, 6) and (6, 3) for example; but is apparently absent in 3-dimensional nets.

Beyond the semi-regular structures lie the Wellsean structures, those in which both the polygonality n and the connectivity p are fractional, in the Schlaefli symbol (n, p). [22] These structures would lie diagonally, to the left or right and displaced vertically, from the entries into the table of Schlaefli symbols for the regular polyhedra, plane nets and 3-dimensional nets. Wells discovered many Wellsean structures, specifically classifying them as possessing fractional polygonality and fractional connectivity,in the course of his exploration of the space of 3-,4-connected nets [22].

It appears that the first actual Wellsean structure was identified, though not topologically indexed as such, by W.L. Bragg and W.H. Zachariasen, from their determination of the crystal structure of the silicate mineral called phenacite, Be_2SiO_4. [23] This was an especially difficult structure determination because of the rhombohedral symmetry ($R\underline{3}$) and the large size of the unit cell (6 molecules of Be_2SiO_4 in each unit cell). In W.L. Bragg's description he reports, "It is difficult to give a clear figure of the structure, because the unit cell is large and a pattern with rhombohedral symmetry is harder to depict than one based upon rectangular axes. The principles of the structure are very simple, however, and are readily traced in a model. It is formed of linked tetrahedra, with Si and Be at their centers. Each O of the independent SiO_4 groups also forms part of 2 neighboring tetrahedra around Be atoms. Thus each Si is surrounded by 4 O atoms, and each Be by 4 O atoms, and each O is linked to 2 Be atoms and one Si atom at the corners of an equilateral triangle." [24]

With the identification of this structure type, which Bragg pointed out could be simplified by replacing the Be and Si atoms by one atom type, to a binary compound of formula A_3B_4 with a smaller unit cell with hexagonal symmetry ($P6_3/m$), entry had been made into a topologically new class of crystal structures. In such a hexagonal lattice, there are four 3-connected vertices and three 4-connected vertices in the asymmetric unit of structure, and there are six 6-sided polygons and four 8-sided polygons in the asymmetric unit of structure [9]. The Schlaefli symbol for the phenacite structure is therefore ($6^{4/5}$, 3.4285….). In 1939 and 1940, Juza and Hahn [25] synthesized and reported the crystal structure of Ge_3N_4, patterned on the structure of phenacite in the manner described by Bragg. [26, 27]. About 17 years later, the synthesis and crystal structure of another polymorph of Ge_3N_4, with nearly an identical density, and the corresponding syntheses of the isomorphous α- and β-Si_3N_4 polymorphs was reported in part by several groups including Hardie and Jack [26] and Ruddleson and Popper [27].

Motivated by the reports of the synthesis of the IV-V nitrides, 1st principles calculations were performed on a hypothetical carbon nitride phase, C_3N_4, patterned on the β-Si_3N_4 structure type [28]. A semi-empirical formula for bulk modulus developed out of this work, predicted the latter nitride would be of a hardness comparable to diamond [28]. Inspection of the glitter structure, a 3-,4-connected net with 6- and 8-sided polygons in its structural pattern, indicates the Schlaefli symbol (7, $3^{1/3}$). The polygonality is an admixture of 6's and 8's which nonetheless has the integer polygonality 7. It is a Wellsean structure. Its polygonality is higher that in the diamond structures (6, 4), and the graphene tessellation (6, 3), and the molecule Buckminsterfullerene ($5^{2/3}$, 3).

In a separate vein, Waser and McClanahan reported [29] the 1st transition metal-containing crystal structure of a 3-,4-connected net, the Pt_3O_4 structure type, in 1951. As it turns out, this is a semi-regular structure (a Catalan network) with the Schlaefli symbol (8, 3.4285....). Here, as in the structures patterned on phenacite, the connectivity appears as a continued fraction. It might be expected that the symmetry of the structure would be low, on the basis of the low topology. Despite the irrationality of the connectivity, the Pt_3O_4 lattice possesses cubic space group symmetry (Pm3n). The IV-V nitrides are of hexagonal symmetry ($P6_3/m$). Topology (regularity) and symmetry are to some degree independent properties of structures. Even so, the well-known structures of graphite and diamond, for example, independent of their high symmetry, possess high topology.

In 1965, Wells [3] reported 2 new semi-regular (Catalan) 3-,4-connected nets; 2 unique nets possessing the same Schlaefli symbol (8, 3.5714.....). Their topological identity is not distinguishable completely by their Schlaefli symbol, as in other cases, and a distinct topological identity is provided through the use of further labels based upon the number of polygons common to each link and vertex, respectively. In fact one net is comprised of square planar connectivity (space group I4/mmm) and the other contains tetrahedral connectivity (space group $I\underline{4}m2$). Both of these nets possessed trigonal planar vertices bonded with 4-connected vertices. Moreover, a 3rd 3-,4-connected net was reported in this paper, this net with the Schlaefli symbol ($8^{1/2}$, $3^{1/3}$). It possesses orthorhombic symmetry (Pmmm). With this work, Wells began his exploration of the 3-,4-connected networks and began identifying Wellsean networks.

Physically, the polygonality is approximately related to the openness of the structure. The connectivity index is approximately related to the closedness of the structure. For example, the glitter structure has a connectivity index of $3^{1/3}$, it is clearly intermediate between graphite at 3, and diamond at 4. From the connectivity index it can be identified topologically as a hybrid of graphite and diamond. Compare to the Buckminsterfullerene molecule, with a polygonality of $5^{2/3}$, it being intermediate between the pentagonal dodecahedron (5, 3) and the graphene tessellation (6, 3). Another useful topological index obtained from the polygonality and the connectivity, is formed by taking the ratio of n and p:

$$l = \frac{n}{p} \quad (4)$$

Where l is a measure of the form of the structure as related to its average polygon size per unit average connectivity. It is a topological index useful in one sense for identifying similarities between regular, semi-regular and Wellsean structures.

Along the principal diagonal of Table 1 are structures in which the polygonality is the same as the connectivity, the topological index l is therefore unity. Such structures have the very highest topology. To the left of the principal diagonal lies the subspace of structures with l indexes less than one, to the right of the principal diagonal lie structures with indexes l greater than one. As a general index of topological relatedness, it is very interesting to note graphite and glitter both possess l indexes close to 2. It is hoped that this indicial similarity can be elucidated in a future communication about graphite and glitter.

Consideration of the work on 3-,4-connected nets initiated with Bragg and Zacharaisen in 1930 on phenacite; followed by Juza and Hahn in 1940 and Ruddleson and Popper, and Jack and Hardie in 1957, on IV-V nitrides; then the theoretical work of A.F. Wells on the 3-,4-connected networks begun in 1965; and most recently the attempts by the group led by Cohen to synthesize C_3N_4 in the 1990's; there is a clear progression towards the synthesis of 3-,4-connected nets of the 2nd period elements. Quite apart from this development, though apparently converging with it, is the descendency of synthetic allotropes of C from those C nets possessing a high topology, to new forms of C which break to lower topologies, the principal example of which is the self assembly of the Buckminsterfullerene molecule [13].

Isomorphous variants of the parent lattice of C atoms, are the III-IV series of compounds which could adopt the glitter structure. The fully metallic band profile of the hypothetical B_2C phase patterned on the glitter structure, has been briefly described [1]. Alternatively, the adjacency of trigonal centers across faces in the unit cell of glitter, suggests a denser structure in which trigonal planar points are transformed into trigonal bipyramidal points. Such a lattice would appear to be the first 4-,5-connected net, and a good model for exploration of 3rd period IV-V structures which have access to expanded octet hybridization (sp^3d) about the Group V elements, for example phases such as SiP_2.

9. The Foundations of a Spiro Quantum Chemistry

With the foregoing analysis of spiroconjugation presented with respect to the 1-dimensional polyspiroquinoid polymer and the 3-dimensional glitter structure, the foundations have been laid for the elucidation of a new type of quantum chemistry, called spiro quantum chemistry. Spiro quantum chemistry complements the historically developed quantum chemistry of linear polyenes, the annulenes, polyhexes, 2D graphene sheets and the fullerenes. Fragments of the polyspiroquinoid polymer, called the spiro[n]quinoids, constitute the spiro quantum chemistry analogs to the linear polyenes. If one joins the spiro[n]quinoids at their ends, one forms the cyclospiro[n]quinoids. Thus the cyclospiro[n]quinoids constitute the spiro quantum chemistry analog to the annulenes of ordinary quantum chemistry. As they possess analogous nodal properties, in the spiro wave functions, to the ordinary π and π^* wave functions of the linear polyenes and the cyclic annulenes, it may be possible to identify an analog of the "4n+2" rule of aromaticity in the cyclospiro[n]quinoids.

In 2 dimensions one has the analog of the graphene sheet reported as the spirographene sheet in [1]. It's band structure shows it to be semi-metallic, in analogy with the ordinary graphene sheet. Fragments of the graphene sheet are termed poly[m,n]hexes, where the indexes m and n indicate the length and width of the polyhex fragment. In analogy to the poly[m,n]hexes there are the spiro[m,n]hexes, where the indexes m and n indicate the numbers of spiro nodes running lengthwise and widthwise in the spiro[m,n]hex fragment. From the band structure of the parent spirographene sheet, it is apparent that the spiro[m,n]hexes may have important electronic or optical properties. One could indeed wrap the spirographene sheet back onto itself and create spirographene cylinders analogous to the graphene cylinders seen in nanotubes.

In fully 3 dimensions one has the synthetic metal glitter, evidently there is no 3-dimensional analog to glitter in ordinary quantum chemistry, which extends to 2D graphene sheets and no further. Glitter, as a synthetic metal, is already interesting from the point of view of its undoped metallic status. That its metallic state can be so tightly linked to the endo-spiroconjugation in it is also remarkable. One could envision fragments of the glitter lattice, analogous to the recently reported diamondoid hydrocarbon fragments of the diamond lattice, called the glitter[m,n,o]enes in which the indexes indicate the numbers of spiro nodes in the fragment along its length, width and height.

Clearly there is a rich area of research accessible to organic chemists, theorists, materials scientists and biologists in spiro quantum chemistry, and the results in this paper are only the beginning. Each class of spiro analogs to the ordinary hydrocarbons of quantum chemistry presents its own, unique synthetic challenges to the organic chemistry community. And certainly there is a wealth of problems for theorists pointed out by this paper. The abundant literature on spiroconjugation, alluded to in [7], suggests its importance in chemistry, biology, materials science and fundamental theory. It is hoped that chemists may be interested enough by the prospects of this manuscript to take up the cause of elucidating the fundamental features of spiro quantum chemistry.

PART II: SYNTHESIS OF GLITTER

ABSTRACT

Structural analysis on carbonaceous samples produced from high temperature-pressure conditions by Palatnik et al. in 1984 indicate the existence of a metallic allotrope of carbon with a diffraction pattern closely matching that of cubic diamond. A structure proposed for this phase by Konyashin et al. and others suggests that the 4 interstitial carbon atoms occupying ½ the tetrahedral holes in the ordinary cubic diamond lattice are vacant in the new structure. This leads to a transformation of the Fd3m, ordinary cubic diamond structure-type, to a simple face centered cubic carbon lattice in space group Fm3m, with a lattice parameter of 3.56 Å, identical with that of cubic diamond. The new structure supports the diffraction evidence accumulated for this so-called "n-diamond" phase, but does not hold up to a first principles total energy optimization at the DFT level of theory for the fcc lattice, which reports the Konyashin et al. structure to be unstable. Here we report an alternative tetragonal carbon structural-type, which we have called "glitter", that explains the observed diffraction pattern of n-diamond reasonably well, and that is stabilized by extensive spiroconjugation in three dimensions leading to a metallic status for the carbonaceous structural-type.

10. DISCOVERY OF METALLIC CARBON

In 2001 Konyashin et al. [30] reported the observation of a metallic modification of carbon from techniques of high temperature-pressure synthesis. These results confirmed earlier studies by Palatnik et al. [31] in 1984. In their work, the scientists synthesized a crystalline carbon material that reproduced the diffraction pattern of diamond, but in addition, there were the presence of symmetry forbidden diamond reflections contained in the diffraction evidence.

Gogotsi et al. [32], also with work reported in 2001, used a reaction-based method centered upon the carbide derived carbon (CDC) strategy developed in their group, and synthesized bulk polycrystalline samples of carbon, which by interrogation with diffraction techniques were seen to produce Bragg reflections consistent with carbon diamond, but in addition the samples were seen to exhibit some symmetry forbidden reflections of diamond, like the (200) reflection of cubic diamond at 1.78 Å. This work was reminiscent of that described by Konyashin et al. [30] and Palatnik et al. [31], in that symmetry forbidden reflections of diamond were consistently observed, along with the usual diamond polytype reflections, in the diffraction analysis of their carbonaceous materials, as produced in this instance from the CDC strategy.

Hirai et al. [33], in research first reported in 1991, catalogued similar results from studies of shock-compressed graphite using a rapid cooling technique. They observed carbon diamond polytype reflections in their heated, shock-compressed carbon samples, but in addition they recorded the (200), (222) and (420) symmetry forbidden Bragg reflections of cubic diamond in their diffraction analysis.

Collectively, the researchers call this diamond-like material "n-diamond" (or γ-carbon) after the resemblance of its diffraction signature to ordinary carbon diamond. Like the other studies on n-diamond, the work of Hirai et al. demonstrated the metallic nature of the carbon materials they produced. Their report will be used as the basis for a comparison of the calculated diffraction pattern of a novel, hypothetical tetragonal allotrope of carbon called "glitter", that has been theoretically characterized by Bucknum et al. previously [34-36], to that of their n-diamond samples, as described below. Finally, Bursill et al. have recorded similar diffraction observations for carbon ion-implanted carbon samples crystallized in a quartz matrix [37].

A structure for the modified diamond form in these studies, the so-called "n-diamond", has been proposed by Palatnik et al. as early as 1984. It would be physically unrealistic to ascribe the symmetry forbidden reflections observed in these studies to ordinary cubic diamond, and in fact they are proposed to originate from a material consisting of a simple face centered cubic (fcc) cell of carbon atoms, in space group Fm3m. Furthermore, in order to explain the diffraction evidence they obtained, the researchers were constrained to define the fcc n-diamond lattice of carbon as possessing the same unit cell edge as does ordinary cubic diamond (Fd3m) at 3.56 Å. The proposed structure of n-diamond and its relationship to the structure of ordinary cubic diamond are shown in Figure 16.

In this proposal, note that the n-diamond structure has a density exactly ½ that of the ordinary cubic diamond lattice, as there are ½ as many C atoms contained in the same cubic cell. The density of their proposed n-diamond structure is therefore 1.78 g/cm^3 (the x-ray density of diamond is 3.56 g/cm^3).

The justification for this face centered cubic (fcc) structure of n-diamond is simply that in the Fm3m space group, where the restrictions of symmetry forbidden reflections present in the Fd3m lattice of carbon are lifted, the resultant diffraction patterns observed to contain symmetry forbidden reflections of Fd3m cubic diamond are readily explained.

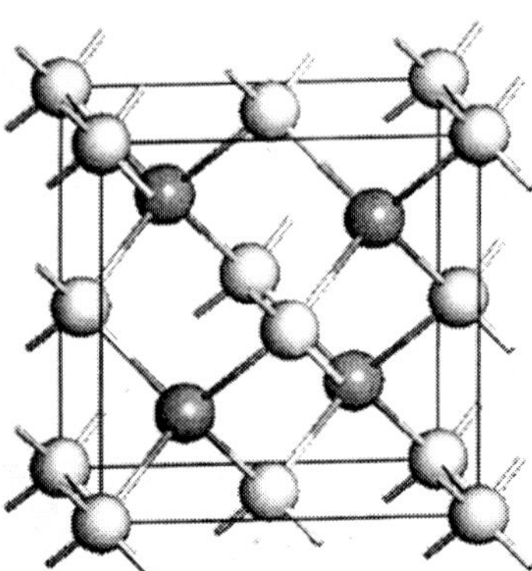

Figure 16. The relationship between Fd3m cubic diamond and an Fm-3m cell.

However, this proposed structure of n-diamond by Palatnik et al. and others seems unlikely to be the true structure from the point of view of elementary chemical bonding theory, as the carbon atoms in the fcc lattice are no longer in the conventional tetrahedral coordination of covalent carbon-carbon single bonding that is ubiquitous in C, as observed in ordinary organic chemistry for example [38]. In the fcc lattice, the C atoms are forced to have a higher coordination number than 4, at distances that are unrealistic for carbon-carbon covalent bonding. In addition, the idea that an allotrope of carbon with ½ the density of cubic diamond could be metallic, is counter-intuitive from a physical standpoint. One would reasonably expect an increase in density from cubic diamond, as corresponding to the onset of metallization in a novel modification of carbon [39].

With the counter-intuitive nature of the proposed model for the structure of n-diamond in mind, the research group of Pickard et al. reported in 2001 [40], in theoretical studies at the state-of-the-art density functional level of theory (DFT) [41], that an Fm3m structure of carbon would be unstable, with the constraint of a cubic lattice parameter of 3.56 Å. They showed that such an fcc C lattice would have an energy-minimized optimum lattice parameter of about 3.08 Å, resulting in a diffraction pattern that would not correspond to the observed evidence for n-diamond. This result from DFT calculations seems to completely rule out the Fm3m structure as an explanation for the observed diffraction pattern of n-diamond. It does not, however, explain the nature of the metallic status observed in these crystalline carbon materials [30-33], nor the unusual diffraction patterns observed for them by several leading research groups around the world, as referenced above.

11. The Glitter Model of Metallic Carbon

Bucknum et al., using a strategy based upon a mixture of sound chemical intuition and semi-empirical molecular orbital calculations (EHMO), to characterize band structures of potential C allotropes [42-45], reasoned that a potential allotrope of carbon might be constructed from a structural basis constituted by a 1,4-cyclohexadieneoid motif. The structural pattern produced by the 1,4-cyclohexadieneoid motif was suggested to them as a potential allotropic structure of C, strictly based upon an analogy with the diamond and graphite polytypes of carbon, which are built upon cyclohexaneoid and benzenoid motifs, respectively. In some sense, such a structure would represent a form of carbon intermediate between graphite and diamond in the degree of unsaturation in the lattice, and in other properties including the coordination number.

Therefore, in 1994 Bucknum et al. reported the prediction of a tetragonal modification of carbon, in space group $P4_2/mmc$ (#131), which contains 6 atoms in the unit cell and has a density of about 3.08 g/cm^3 (the density of such a lattice of C could be greater than 3.2 g/cm^3 by employing a corresponding configuration of the tetragonal lattice parameters). This calculated density is based upon the optimized lattice parameters for the tetragonal cell given as **a** = 2.560 Å and **c** = 5.925 Å. These lattice parameters have been optimized using a computational procedure based upon Density Functional Theory (DFT) [41], the computational procedure is known as CASTEP and it was developed at Cambridge University by Payne et al. [46]. The optimization calculations were performed in CASTEP using a functional based upon the local density approximation (LDA). The earlier optimization calculations on the fcc carbon cell by Pickard et al. were performed with this code as well [40, 46].

The tetragonal structure-type was built entirely upon a 1,4-cyclohexadieneoid motif [47-49], as shown in Figure 17, and constituted a 3-,4-connected network of C, related to many other such 3-,4-connected structural-types that were derived and constructed by A.F. Wells in the period from the early 1950's to the mid 1980's [50-51]. The structure was given the name "glitter" based upon the metallic nature of the bonding in it, produced from extensive p_σ and p_{spiro} interactions within the network [34-38].

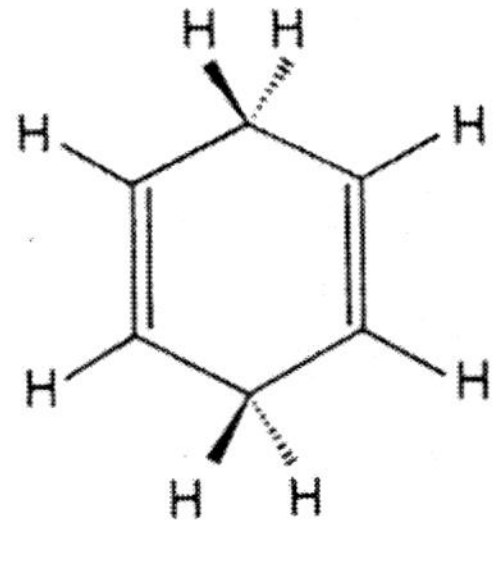

Figure 17. The 1,4-cyclohexadiene model molecule.

In their initial report on the glitter structure, Bucknum et al. showed the material was alternatively built upon a motif of 3-,4-connected 1-dimensional polymer chains of C, with the adjacent 1-dimensional chains of such substructures coordinated to each other in an orthogonal fashion. The resulting glitter allotrope, in the tetragonal space group $P4_2/mmc$, was thus shown to be intermediate in constitution to the polyethylene chains of the diamond lattice, in which adjacent chains are joined orthogonally (and with the latter structure in space group Fd3m), and the polyacetylene chains of the Wells-constructed 3-connected net known as (10, 3)-b, in which adjacent polyacetylene chains are joined orthogonally (and with the latter structure in space group $I4_1/amd$) [50-52]. In addition, it was seen that glitter could be derived from the known mineral structure of Cooperite [53] by replacing the square planar vertices in the Cooperite structural-type with the trigonal planar atom pairs of glitter. Such a transformation is described as a topological isomorphism. A drawing of the glitter unit of pattern is shown in Figure 18 and Figure 19 shows a view of an array of several unit cells of glitter.

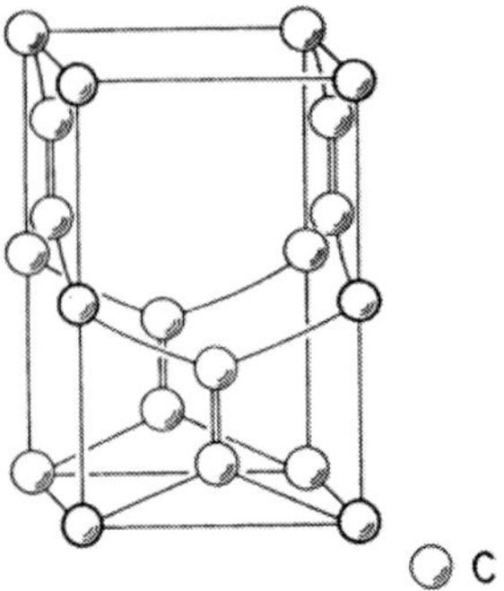

Figure 18. The unit of pattern of glitter.

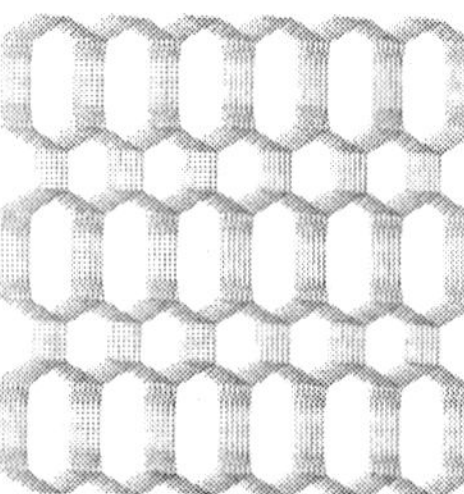

Figure 19. An extended view of glitter.

By calculating a theoretical diffraction pattern for the glitter lattice, based upon, for example, the DFT optimized set of lattice parameters given above as **a** = 2.560 Å and **c** = 5.925 Å, one can show a good fit to the diffraction pattern of n-diamond, as reported, for example, by Hirai et al. in 1991 [33]. The close match between the calculated Bragg reflections for the glitter lattice and those reflections measured for the n-diamond material, suggest the central message of this communication. Perhaps the structure of n-diamond and the structure of glitter are one and the same. A tabulation of the associated diffraction data is given in Table 2 [33, 54]. Note that the mean of the deviations of the 10 matched Bragg reflections in Table 2 is given as x(mean) = 0.001983 nm/reflection (0.01983 Å/reflection) and the standard deviation of the corresponding dataset for the 10 matched Bragg reflections is given as σ_x = 0.002363 nm/reflection (0.02363 Å/reflection). Therefore, all the data fit within $3\sigma_x$ of x(mean) (0.009072 nm/reflection) with the glitter(003)-to-n-diamond(111) pair of matched reflections having the largest deviation at 0.008500 nm (0.08500 Å).

With the data in Table 2, which, as stated above, consist of 10 n-diamond reflections, as well as the calculated tetragonal glitter and ordinary cubic diamond reflections, it can be seen that the set of data is entirely consistent with that of Fd3m cubic diamond, except that 3 of the reflections, the (200) at 1.780 Å, the (222) at 1.040 Å, and the (420) at 0.7960 Å, are symmetry forbidden cubic diamond reflections in the analysis. Quite in contrast, one can consider the data from the theoretical diffraction pattern of glitter, where it can be seen that all 10 reflections of n-diamond are reproduced in the glitter spectrum, with only the (003) glitter reflection, at 1.975 Å, being symmetry forbidden. The fit between glitter and the n-diamond data is thus seen to be a closer fit than that of cubic Fd3m diamond, especially given that the lattice parameters giving rise to this glitter diffraction pattern have been optimized for the lattice at the DFT level of theory [41, 46].

Table 2. Observed diffraction data of n-diamond compared to theoretical diffraction data of $P4_2/mmc$ tetragonal glitter and Fd3m cubic diamond

calculated glitter reflections a = 0.2560 nm, c = 0.5925 nm (hkl)	d-spacing, nm	n-diamond reflections* d-spacing, nm	cubic diamond reflections** d-spacing, nm
100	0.2560		
001	0.5925		
110	0.1810	0.1780 (200)	"forbidden"
101	0.2350		
111	0.1731		
200	0.1280		
002	0.2963		
102	0.1937		
120	0.1145		
201	0.1251	0.1260 (220)	0.1261 (220)
211	0.1124		
221	0.08947	0.08980 (400)	0.08916 (400)
212	0.1068	0.1040 (222)	"forbidden"
222	0.08656		
300	0.08533		
003	0.1975	0.2060 (111)	0.2060 (111)
103	0.1564		
130	0.08095	0.07960 (420)	"forbidden"
301	0.08446		
311	0.08021		
331	0.06003		
313	0.07491		
333	0.05771		
203	0.1074	0.1070 (311)	0.10750 (311)
302	0.08200	0.08180 (331)	0.08182 (331)
320	0.07100	0.07260 (422)	0.07281 (422)
223	0.08228		
232	0.06905	0.06830 (511)	0.06864 (511)
332	0.05912		
323	0.06682		
321	0.07050		
312	0.07809		
213	0.09905		
104	0.1282		
401	0.06363		

*H. Hirai and K. Kondo, Science, 253, 772 (1991)
** JCPDS6, Card # 675

Yet a second piece of evidence supporting the glitter model as an explanation of the structure of n-diamond, is the data supplied from a state of the art, density functional theory (DFT) calculation of the band structure of the corresponding tetragonal lattice [55]. In fact, the structure is shown to be a good metal at the DFT level of theory, with the π^* band dipping down into the occupied bands of glitter at symmetry point M in the reciprocal space of the material [34-36]. Metallic status for glitter is consistent with the observations of Palatnik et al., and others, mentioned above, with regard to the electrical conductivity of n-diamond [30-

33]. The calculated band structure of glitter, with the lattice parameters close to those given above, is shown in Figure 20.

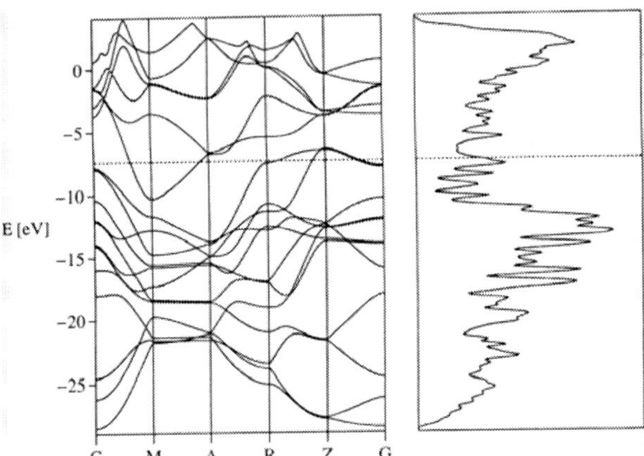

Figure 20. Calculated band structure and density of states (DOS) of tetragonal glitter at DFT level of theory.

12. Theoretical Stability of Glitter

First principles calculations of the stability of the glitter lattice have been performed with the CASTEP computational method described above. It has been shown that under the local density approximation (LDA) the DFT-based CASTEP method calculates a stability of 0.511 eV/C atom above the corresponding stability of graphite, which is the zero on the energy scale [46]. As a comparison, fullerene is calculated to be 0.3 eV/C atom less stable than graphite by the same method. Although this is a fairly large energy above that of graphite, if the glitter structure was synthesized it would persist, as kinetic routes to its decomposition to graphite or diamond have high activation barriers due to the reconstructive nature of such a phase transition [48]. These considerations are based upon the unique 3-,4-connected bonding in the structure.

In addition to the kinetic parameters, the spiroconjugation effects reported for it previously [56], lend great support for the relative stability of glitter as a structure of carbon. Conjugation and resonance are known to be important themes in the explanation of the reactions of organic chemistry, and in explaining the structural stability of graphite and the fullerenes [38, 53]. In the graphene sheet, one is constrained to conjugation in the 2D plane. With the fullerenes, one is constrained to conjugation on a 2D spherical surface. From the MO calculations done on glitter in the 1990’s [34-36], it is clear that another type of p-orbital conjugation, called spiroconjugation [57-70], is present in the glitter lattice in fully 3D. It constitutes the first fully 3-dimensional conjugated organic system that has ever been conceived in the chemical literature [34-36].

In direct analogy with the conjugation present in the graphene sheet, as shown in Figure 21 [53], there are a set of resonance structures that can be written over the glitter unit cell to indicate its relative stability from spiroconjugation, this is illustrated in Figure 22.

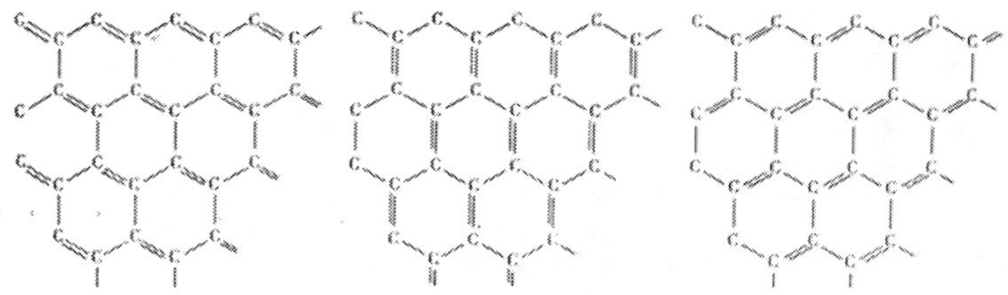

Figure 21. Resonance structures of graphite in 2D.

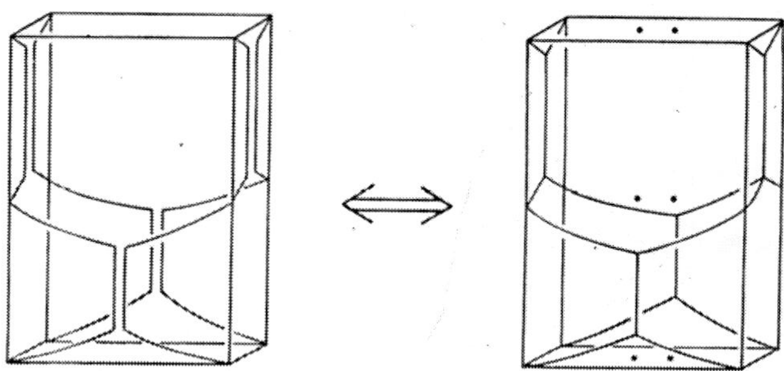

Figure 22. Resonance structures of glitter in 3D.

It is clear from these resonance structures in 3D, that the electrical conductivity in glitter will be highly anisotropic, with conduction electrons, as π electrons carried by the overlapping π and π* bands in glitter's electronic structure, moving in pairs in a direction parallel to the **c**-axis of the unit cell. It is indeed a very unusual, organic metal with a correspondingly unusual π electron pair conductivity mechanism, and potentially a unique type of stability due to the organic chemical phenomenon of conjugation, as spiroconjugation [57-70], exhibited in 3D in the electronic structure of glitter [34-36].

In comparison, one can consider the various members of the fullerene family, the parent Buckminsterfullerene molecule C_{60}, although calculated to be 0.3 eV/C atom less stable than graphite [77], was predicted by Osawa in 1970 [100-101] to be a uniquely stable molecule based upon the resonance structures that could be drawn over the C_{60} skelton. It wasn't until 15 years later in 1985 that Kroto et al. [73] accidently stumbled onto the existence and relative stability of the fullerene family (all of which have resonance forms which can be written over their cage structures) from a synthesis based upon laser ablation of a graphite disk, followed by expansion of the resultant carbon species through a supersonic nozzle with He carrier gas. The work of Kroto et al. therefore confirmed the prediction of closed shell carbon molecules by Osawa and others [71-72].

It can be seen from this lesson that conjugation and resonance are a hallmark of stability in molecules and extended structures of carbon. One should therefore take seriously the proposal of glitter as an extended structure of carbon, even in addition to the close match between its calculated diffraction pattern as compared to that of the experimentally observed n-diamond diffraction data, because it, like its sibling structures graphite and the fullerenes, has a unique stability due to the spiroconjugation [57-70] present in fully 3D within the structure [34-36].

The symmetry adapted π and π* crystal orbitals of glitter, at symmetry point M of the Brillouin zone of the tetragonal material, shown in Figure 23, are seen to support the picture of resonance in 3D shown in Figure 7. However, this comparison should be looked upon with the caveat that the glitter crystal orbitals are derived from a molecular orbital (MO) perspective, while the valence bond (VB) method is employed in arriving at the resonance structures of glitter shown in Figure 7 [34-36].

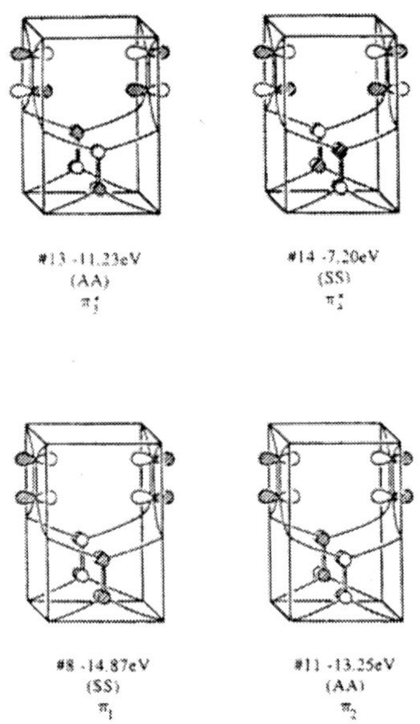

Figure 23. The π and π* crystal orbitals of glitter.

13. Conclusions

It is quite evident from the data comparison shown in Section 2, in which the mean deviation over 10 matched Bragg reflections between glitter and n-diamond is less than 0.02 Å/reflection, that glitter is a plausible model for n-diamond. Of course, there are many proposals for carbon allotropes that have been published [74-100], some of these are chemically intuitive, like the original fullerene proposal by Osawa [71-72], while other models for carbon allotropes, like the fcc structure of carbon proposed for n-diamond, are counterintuitive. In what way can we eliminate these myriad other possible carbon allotropes as structures of n-diamond?

Clearly some of these hypothetical carbon allotropes have diffraction patterns which show reflections that may overlap with those observed for n-diamond. However, the possibility that the complete set of 10 n-diamond reflections observed by Hirai et al. [33] should be entirely reproduced by another hypothetical structure of carbon, in addition to the glitter structure, is remote. The unlikelihood of such a set of 10 coincidental overlaps is even more remote, when one includes in the analysis that the proposed carbon structure should be chemically intuitive, as well as unique in its electronic structure from the point of view of conjugation and resonance effects, as well as being metallic in its band profile.

Glitter is special in that it satisfies all of these requirements. It reproduces the n-diamond diffraction spectrum reasonably well, it is a chemically intuitive structure based upon reasonable bond parameters (the carbon-carbon single bonds in this optimization are 1.526 Å and the carbon-carbon double bonds are 1.330 Å, while the C-C=C angle is 123 degrees and the C-C-C bond angle is 114 degrees), it is comprised of C atoms in archetypal tetrahedral and trigonal connection motifs [38], the trigonal carbon atoms are grouped into parallel pi bonded pairs rather than existing as separate radical centers or being in orthogonally coordinated pairs, it has the unique property of conjugation and resonance, and it is a metallic carbon structure.

Lastly one must consider the myriad of reflections calculated for glitter but not observed in the n-diamond analysis. Obviously, from considerations of both powder diffraction and Laue diffraction, all Bragg (Laue) reflections possible from a given material are rarely observed in practice. There are many reasons for diminished diffracted x-ray intensity, the most important of which is symmetry [101]. In the analysis given in Section 2, the glitter reflection (003) at 1.975 Å is calculated to be forbidden by symmetry [54]. Therefore its potential match to the n-diamond reflection at 2.060 Å is in question.

It is possible that by extending the degree of reciprocal space that is sampled in an n-diamond diffraction experiment, and lengthening the exposure time over which n-diamond diffraction data is collected, one might see some additional reflections as predicted by the data given in Table 2. As n-diamond is a phase produced only intermittently, and therefore n-diamond has not been exhaustively explored to date, it is not surprising that a complete diffraction pattern has not been obtained yet. Of course, for the reasons mentioned above, the glitter model is still a very plausible structure for n-diamond, and those reflections predicted for it in Table 2 should be taken seriously by investigators in the field of carbon materials research.

As to its utility, polycrystalline glitter will be potentially useful as an abrading material for machining applications and, in the form of a single crystal, as a material that exceeds even cubic diamond in its degree of hardness [102-105]. That the material is an organic metal is yet another technological bonus of this structure. Potentially it could be used in situations where strength and conductivity are both required characteristics. The use of glitter as a single crystal anvil material in an opposed anvil high pressure device, analogous to the well-known diamond anvil cell (DAC) [106] is obvious. Overall, it seems that there is quite a lot of technological promise in the glitter structure with its metallic and superhard characteristics. Finally we should mention the synthetic work on glitter and its isovalent analog silicon dicarbide being carried out in Romania contemporaneously with the publication of this manuscript, indeed Stamatin et al. have shown that these glitter phases can be readily synthesized from Novolac resin (a Bakelite resin) using a cross-linking agent like HMTA (hexamethylene tetramine), which is appropriate to cross-linking in a spiroconjugated fashion [107].

PART III: CLASSIFICATION OF CARBON ALLOTROPES AND GRAPHS

ABSTRACT

In this section, we describe the tenets of a chemical topology of crystalline matter and certain associated rational approximations to the transcendental mathematical constants ϕ, e and π, that arise out of considerations of both: (1) the Euler relation for the division of the sphere into vertices, V, faces, F, and edges, E, and: (2) its simple algebraic transformation into the so-called Schläfli relation, which is an equivalent mathematical statement for the polyhedra, in terms of parameters known as the polygonality, defined as n = 2E/F, and the connectivivty, defined as p = 2E/V. It is thus the transformation to the Schläfli relation from the Euler relation, in particular, that enables one to move from a simple heuristic mapping of the polyhedra in the space of V, F and E, into a corresponding heuristic mapping into Schläfli-space, the space circumscribed by the parameters of n and p. It is also true, that this latter transformation equation, the Schläfli relation, applies only directly to the polyhedra, again, with their corresponding Schläfli symbols (n, p), but as a bonus, there is a direct 1-to-1 mapping result for the polyhedra, that can be seen to also be extendable to the tessellations in 2-dimensions, and the networks in 3-dimensions, in terms of coordinates in a 2-dimensional Cartesian grid, represented as the Schläfli symbols (n, p), as discussed above, which do not involve rigorous solutions to the Schläfli relation. For while one could never identify the triplet set of integers (V, F, E) for the tessellations and networks, that would fit as a rational solution within the Euler relation, it is in fact possible for one to identify the corresponding values of the ordered pair (n, p) for any tessellation or network. The identification of the Schläfli symbol (n, p) for the tessellations and networks emerges from the formulation of its so-called Well's point symbol, through the proper translation of that Well's point symbol into an equivalent and unambiguous Schläfli symbol (n, p) for a given tessellation or network, as has been shown by Bucknum et al. previously. What we report in this communication, are the computations of some, certain Schläfli symbols (n, p) for the so-called Waserite (also called platinate, Pt_3O_4, a 3-,4-connected cubic pattern), Moravia (A_3B_8, a 3-,8-connected cubic pattern) and Kentuckia (ABC_2, a 4-,6-,8-connected tetragonal pattern) networks, and some topological descriptors of other relevant structures. It is thus seen, that the computations of the polygonality and connectivity indexes, n and p, that are found as a consequence of identifying the Schläfli symbols for these relatively simple networks, lead to simple and direct connections to certain rational approximations to the transcendental mathematical constants ϕ, e and π, that, to the author's knowledge, have not been identified previously. Such rational approximations lead to elementary and straightforward methods to estimate these mathematical constants to an accuracy of better than 99 parts in 100.

14. A Chemical Topology

Bucknum [108] in work first described in 1997, outlined a general scheme for the systematic classification and mapping of the polyhedra, 2-dimensional tessellations and 3-dimensional networks in a self-consistent topological space for these structures. This general scheme begins with a consideration of the Euler relation [109] for the polyhedra, shown below as Equation 5, which was first proposed in 1758 to the Russian Academy by Euler, and was, in fact, the point of departure for Euler into a new area of mathematics thereafter known explicitly as topology.

$$V - E + F = 2 \tag{5}$$

Relation (5) stipulates that for any of the innumerable polyhedra, the combination of the number of vertices, V, minus the number of edges, E, plus the number of faces, F, resulting from any such division of the sphere, will invariantly be that number 2, known as the Euler characteristic of the sphere. The variables known as V, E and F are topological properties of the polyhedra, or, in other words, they are invariants of the polyhedra under any kind of geometrical distortions. It is from this simple Eulerian relation, that we can develop a systematic and, indeed otherwise rigorous, mapping of the various, innumerable structures that present themselves, in levels of approximation, as models for the structure of the real material world within the domain of that area of science known as crystallography.

About a century after Euler's relation for the polyhedra was first proposed, as described above, the German mathematician Schläfli introduced a simple algebraic transformation of Euler's relation, for various purposes of understanding the relation better, and adopting it more effectively in proofs [110]. Thus, Schläfli introduced two new topological variables, like V, E and F before them, that were derived from them. Schläfli, therefore, defined the so-called "polygonality", hereafter represented by n, of a polyhedron as the averaged number of sides, or edges, circumscribing the faces of a polyhedron. He conveniently defined such a polygonality, as n = 2E/F, where, in this instance one can see that because each edge E straddles two faces, F, the definition is rigorous. Similarly, Schläfli introduced the topological parameter called the "connectivity", hereafter represented by p, of a polyhedron as the averaged number of sides, or edges, terminating at each vertex of a polyhedron. He conveniently defined such a connectivity, as p = 2E/V, where, in this instance one can see that because each edge E terminates at two vertices, V, the definition is rigorous.

From these definitions of n and p as topological parameters of the polyhedra, Schläfli was able to show quite straightforwardly, by algebraic substitution, that a further relation exists among the polyhedra in terms of their Schläfli symbols (n, p). [110] It is from this equation, the Schläfli relation, shown as Equation 6 below, that one can see that not only do the polyhedra rigorously obey 6, but it is also true that their indices as (n, p), that serve as solutions to 6, in addition lead to a convenient 2-dimensional grid, or Schläfli space, over which the various polyhedra can be unambiguously mapped, as has been explained by Wells in his important 1977 monograph on the subject [111].

$$\frac{1}{n} - \frac{1}{2} + \frac{1}{p} = \frac{1}{E} \qquad (6)$$

Figure 24 due to Wells [111], below, illustrates the application of this type of Schläfli mapping for the regular Platonic polyhedra, where one sees that the point that belongs to the origin of this mapping, is indeed given by the Schläfli symbol (n, p) = (3, 3). The symbol (3, 3) represents the Platonic solid known as the tetrahedron, or by the symbol "t" in the map, known since Antiquity by the Greeks. Thus, as it is cast as the origin of this mapping of polyhedra, it is apparently, the only self-dual polyhedron. Similarly (4, 3) is the Platonic solid of the Greeks known as the cube, or "c" in the map, (5, 3) is the Platonic solid of the Greeks known as the dodecahedron, or "d" in the map, (3, 4) is the Platonic solid of the Greeks known as the octahedron, or "o" in the map, and (3, 5) is the Platonic solid of the Greeks known as the icosahedron, or "i" in the map.

Although the mapping of the Platonic polyhedra, shown in Figure 24, indeed involves only those polyhedra in which the ordered pairs (n, p) are integers, it is readily transparent that one could *magnify* the map to include those polyhedra in which the polygonality "n" is fractional, these are the so-called Archimedean polyhedra, discovered by Archimedes in Ancient Greece [112].

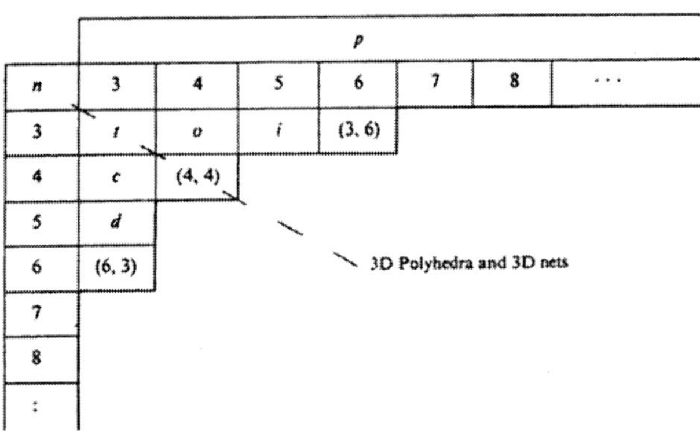

Figure 24. Topology mapping of the Platonic polyhedra due to Wells.

In addition, the map could be *magnified* to include those polyhedra in which the connectivity "p" is fractional, these are the so-called Catalan polyhedra, discovered in Europe in the 19th century [113]. And finally, the so-called Wellean polyhedra [114], discovered in the 21st century [115], in which both indexes, "n" and "p", are fractional, could be mapped in this Schläfli space of the polyhedra, without any loss of mathematical rigor.

15. Mapping of Tessellations and Networks in Schläfli Space

Wells also suggested [111], that the tessellations, which are structures or tilings extended into 2-dimensions and filling the plane, could have nominal labels attached to them, in the form of the Schläfli symbols (n, p), that, while not leading to rational solutions of the Schläfli relation, were, still it seems, rigorously defined from inspection of the topology of these elementary tessellations. Thus in order to *extend* the mapping in Schläfli space, the space of (n, p), some loss of rigor with regard to the Schläfli relation for the polyhedra had to be introduced, when considering the patterns known as tessellations. Wells, therefore included, explicity, the mapping of the square grid, given by the Schläfli symbol (4, 4), and the honeycomb grid, given by (6, 3), as well as the closest-packed tessellation, given by (3, 6), in his topology mapping, shown in Figure 24, along with the Platonic polyhedra. He thereby *extended* the mapping to the tessellations, and later he implied that such a mapping could be *extended* to include the 3-dimensional (3D) networks as well, with a concomitant further loss of mathematical rigor, in that the values assigned as (n, p) to the various tessellations and networks were not rigorous solutions to Equation 6.

It is also true that Wells [114], perfectly well introduced a systematic and rigorous coding of the topology of tessellations and networks he worked with, which is now called the Wells point symbol notation, and that this was a simple coding scheme over the circuitry and valences, about the vertices, in the unit of pattern of the tessellations and networks. The Wells point symbol notation was, however, nonetheless an important development for the rigorous mathematical basis it put the tessellations and networks on, formally, as quasi-solutions (n, p) for the Schläfli relation shown as Equation 6.

Therefore, in a generic case of the Wells point symbol notation, one could have such a symbol for an a-, b-connected, binary network, given as $(A^a)_X(B^b)_Y$, such that the exponents "a" and "b" nominally represent the valences of the 2 vertices in the tessellation or network of interest, and the bases "A" and "B" give the respective relative polygon sizes (circuit sizes) in the tessellation or network, while the parameters "X" and "Y" describe the binary stoichiometry of the network. In this generic case, we see that "X/Y" represents the number of structural components, identified by their topology character as (A^a), to the number of structural components identified by their respective topology as (B^b), that occur in this characteristic ratio in the structure, as specified by the unit of pattern.

Despite his invention of this elegant notation, Wells, for some odd reason, never explicitly showed how to *translate* the language of the Wells point symbol $(A^a)_X(B^b)_Y$, rigorously into a Schläfli symbol (n, p), as was later shown elsewhere. Thus Bucknum et al., in 2004 [116], showed that the translation of the Wells point symbol into an, otherwise, from the perspective of Equation 6, rigorous set of values (n, p) for the purpose of mapping tessellations and networks, was achievable if one used the following straightforward, simple formulas (applicable in this case for a generic Wellsean, binary stoichiometry structure) for the ordered pair (n, p), which can later be employed in the mapping of the structure, as is shown by Equations 7.

$$n = (a \cdot A \cdot X + b \cdot B \cdot Y)/(a \cdot X + b \cdot Y) \tag{7a}$$

$$p = (a \cdot X + b \cdot Y)/(X + Y) \tag{7b}$$

It should, of course, be noted that one has to proceed with care in using Equations (7a & 7b), in this topology analysis of structures, by carefully normalizing the circuitry traced around the p-connected vertices of the structure (paying careful attention to the parameters "a" and "b" above), by rigorously translating the circuitry of a given structure, into a vertex connectivity for the network of interest, by employing a vertex translation table like that shown below.

Table 3. Vertex Connectivity, p, as a Function of Circuit Number

name	vertex connectivity	circuit number
trigonal planar	3	3
square planar	4	4
tetrahedral	4	6
trigonal bipyramidal	5	9
square pyramidal	5	10
octahedral	6	12
cube centered	8	24
anti-cube centered	8	28
closest packed	12	60

Therefore, from the use of Equations 7, for a binary stoichiometry, Wellsean net with topology $(A^a)_X(B^b)_Y$, or some other homologous translation formulas, as shown, for example, by Bucknum et al. for various elementary structural cases [116], it becomes, evidently, rigorously possible to *precisely* map the topology of any structure, including of course the polyhedra, but *extendable* to the vast body of the known tessellations and networks, that have been discovered and characterized crystallographically, otherwise by their symmetry character, now by their topological character in the form of a mapping in an *extended* Schläfli space, as is shown in Figure 25 below.

n \ p	3	4	5	6	7	8	...
3	t	o	i	(3,6)	(3,7)	(3,8)	
4	c	(4,4)	(4,5)	(4,6)	(4,7)	(4,8)	
5	d	(5,4)	(5,5)	(5,6)	(5,7)	(5,8)	
6	(6,3)	(6,4)	(6,5)	(6,6)	(6,7)	(6,8)	
7	(7,3)	(7,4)	(7,5)	(7,6)	(7,7)	(7,8)	
8	(8,3)	(8,4)	(8,5)	(8,6)	(8,7)	(8,8)	
⋮							

Figure 25. Extended Schläfli space of the Platonic Structures.

16. Survey of Mapped Patterns and Carbon Allotropes

As Equations 5 & 6, and Figure 24 & 25 explicitly reveal, it is the Platonic solids that form the basis of this mapping formulation of structures described in this paper. These forms, as shown in Figure 26 with their appropriate polyhedral face symbolism [117], as discovered in Ancient Greece from the application of pure thought, were implicated later on in Plato's Timaeus, as the building blocks of Nature [118]. With the advent of modern crystallographic techniques by the Bragg's in the 20th century [119], we have come to learn that the structure of matter does indeed often take on various vestiges of these eternal objects. And so they have come to be important in modern structural chemistry as elucidated by Pauling [120] and others.

The polyhedra, thus forming the basis of the topology map of structures in Figure 25, and also rigorously obeying the topology relations shown as Equations 5 & 6 above, are positioned uniquely in this construction to support the vast space of tessellations and networks that, as we have seen in the preceding Section, can be mapped, rigorously, in Figure 25 by the identification and proper translation of their Well's point symbols, as described above, into ordered pairs as Schläfli symbols (n, p). Plato's great work, Timaeus, thus predicted the ascendancy of the material world into perfect forms, in which the Platonic polyhedra hold primacy and support the overall organizational structure of matter, from which the innumerable other polyhedral objects, and the innumerable 2D tessellations, and the innumerable 3D networks, together all emerge as perfect objects, in this scheme.

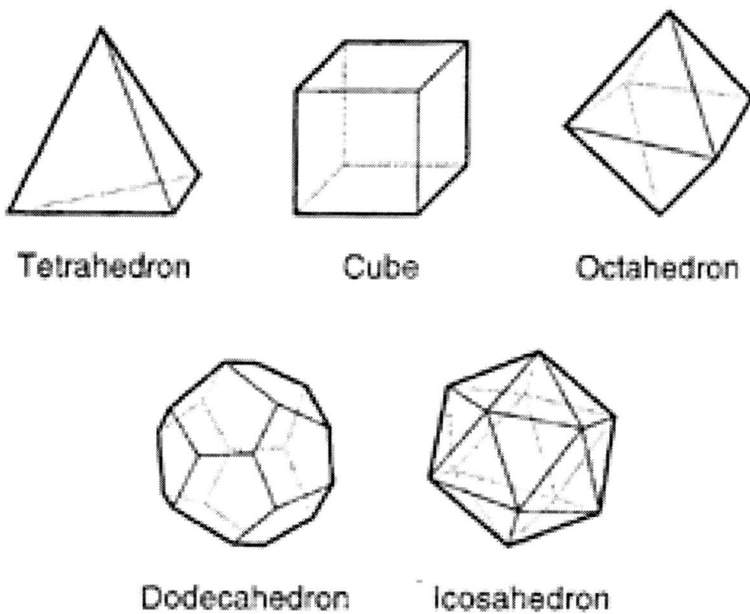

Figure 26. The Platonic polyhedra, with their corresponding polyhedral face symbols and Wells point symbols, comprised of the tetrahedron (34 and 33), the octahedron (38 and 34), the icosahedron (320 and 35), the cube (46 and 43), and the dodecahedron (512 and 53).

Later, it has been shown by Duchowicz et al. [121], that indeed molecular structures (graphs) can be represented in the scheme of Figure 25, and they have a corresponding set of two topology relations (in addition to n = 2E/F and p = 2E/V), shown as Equations 8 & 9 below, that govern their mapping into Figure 25.

$$V - E + F = 1 \qquad (8)$$

$$\frac{1}{n} - \frac{1}{2} + \frac{1}{p} = \frac{1}{2E} \qquad (9)$$

In this way, one can see that the overall chemical topology scheme described here, is a complete description of the topology of all matter constituting the material world. This communication will not treat the specific applications of Equations 8 & 9, but it will be left to the reader to refer to those applications suggested in the literature [121].

Moving from the polyhedra and molecular fragments, as described above, one can then map the regular tessellations as shown in Figure 27, with the honeycomb tessellation, given by the Wells point symbol notation as 6^3, and translated into the mapping symbol or Schläfli symbol as (n, p) = (6, 3), and the square grid, given by the Wells point symbol notation as 4^4, and translated into the mapping symbol or Schläfli symbol as (n, p) = (4, 4), and finally, the third regular tessellation, which thus outlines the space of Figure 2 in terms of the tessellations, as the closest-packed grid, given by the Wells point symbol notation as 3^6, and translated into the mapping symbol or Schläfli symbol as (n, p) = (6, 3).

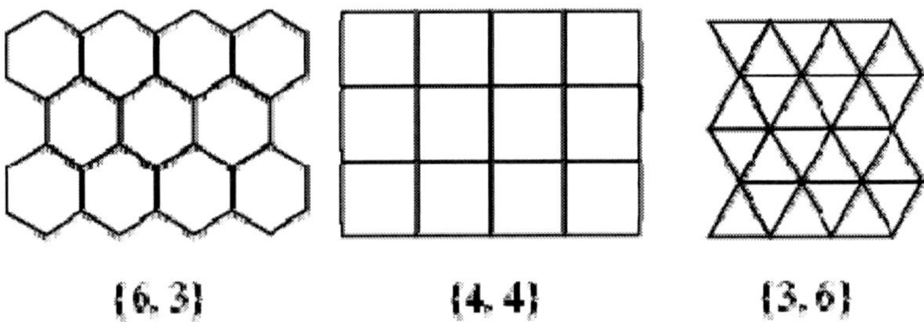

Figure 27. The Platonic tessellations, with their corresponding Wells point symbols, given as the closest-packed grid (36), the square grid (44) and the honeycomb grid (63).

One can, of course, insert all manner of hybrid tessellations, among these 3 regular ones, and generate innumerable Archimedean, Catalan and Wellsean tessellations. Some hybrid tessellations of the square grid-honeycomb grid pair have been analyzed topologically by Bucknum et al. [122]. There are, of course, an infinity of such structural tessellations, and they indeed fill the space in the neighborhood of the borderline between the polyhedra and the tessellations, on the one hand, and the tessellations and the networks in 3D, on the other hand. It is also true that Wells and others [111], have identified tessellations of the plane comprised of 5-gons & 7-gons, and their have been tessellations of 4-gons & 6-gons that admit unstrained 8-gons, and there are many more tessellations proposed, some of which have been taken as models of various C allotropes [123-125], which essentially can possess any n-gons in their pattern, provided that the restraint of being regular n-gons is relaxed. And it is thus true, that all of this infinity of tessellations can be mapped, rigorously, by the methods outlined above.

Finally, the map in Figure 25 outlines the 3D networks, and a prominent member is, of course, the diamond lattice given by the Wells point symbol 6^6, which is translated [116] into the Schläfli symbol (6, 4). By examination of Figure 25, one can see that the diamond network, given the Schläfli symbol (6, 4), is situated just across the borderline from the 2-dimensional honeycomb tessellation given by (6, 3) in the map. One member of the diamond network topology, is in a cubic symmetry space group of Fd-3m, space group #227, one of the highest symmetry space group patterns.

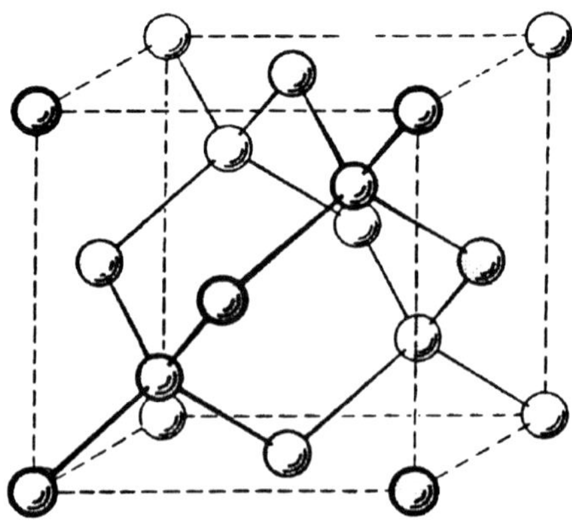

Figure 28. Cubic diamond (3C) polytype, with the Wells point symbol (66), lying in symmetry space group (Fd-3m).

There are, in fact, innumerable possible polytypic patterns within the diamond topology, several of these have been discussed recently by Wen et al. [126] in some detail, and all of them collectively possess the same Wells point symbol of 6^6, and the corresponding Schläfli symbol of (6, 4). It is only by their symmetry character, that the members of the diamond polytypic series can be distinguished from each other. Thus, the simplest cubic diamond polytype, known as 3C, is shown in Figure 28 below.

Thus in the diamond network, which corresponds to the Platonic (integer) topology of the Platonic polyhedra, one can readily trace the uniform 6-gon, puckered circuitry of the network connected together by all 4-connected, tetrahedral vertices. Diamond's topology classifies the network as a regular, Platonic structure-type.

Next, we move to the space between (6, 3) and (6, 4), seemingly between 2D and 3D forms, and investigate what potential structures might emerge along this boundary area. Such an examination turns up two distinct families of Catalan networks, that together possess the Catalan Wells point symbol $(6^6)_x(6^3)_y$. One can see, through this Wells point symbol notation, that we are describing hybrid structures of the honeycomb tessellation, the so-called graphene grid, 6^3, and the diamond network, 6^6. The notation "y/x" specifies the stoichiometry of the net, in terms of the ratio of 3-connected, trigonal planar vertices, to 4-connected, tetrahedral vertices in the hybrid structure [127].

Thus one example of such a class of hybrid "graphene-diamond" structures is shown in Figure 29, and these forms are known, by their hybrid topology, as the "graphite-diamond hybrids". They come in infinite series', in each of two varieties that are known as the ortho-form, and the para-form, these have been elegantly described by Balaban et al. in 1994 [128]. Collectively, as a family, they possess the Schläfli symbol of $(6, 3^{(x/(x+y))})$, where the parameters "x" and "y" have the stoichiometric significance ascribed to them in the preceding paragraph. These structures, as a family, occupy the border-line area between (6, 3) and (6, 4) in the Schläfli map in Figure 25.

Yet another family of "graphite-diamond" hybrid structures to be considered, with the Wells point symbol $(6^6)_x(6^3)_y$, and the corresponding Schläfli symbol given by $(6, 3^{(x/(x+y))})$, is the family of structures described first by Karfunkel et al. [129] in 1992, as being built from the barrelene hydrocarbon molecular fragment, and extended by the insertion of benzene-like tiles to the parent framework, to generate many infinities of derived structures, which all, collectively, possess the hybrid graphite-diamond topology described above. Later, in 2001, Bucknum [130] clarified the details of the parent such structure derived by Karfunkel et al., and he called this structure "hexagonite" and the derived, such structures were known as the "expanded hexagonites". This name was assigned due to the symmetry space group of the parent structure, in P6/mmm, space group #191, and also due to the topology of the family of such structures, in which all circuitry over all members of the family, are comprised of 6-gons. The topological analysis of the hexagonite family, suggests that they begin with the Schläfli symbol (n, p) = $(6, 3^{2/5})$, and extend from there, in descending, discrete increments of the connectivity, p, towards their termination at (n, p) = (6, 3), in the limit of the graphene grid topology. The parent "hexagonite" is shown in two views in Figure 30.

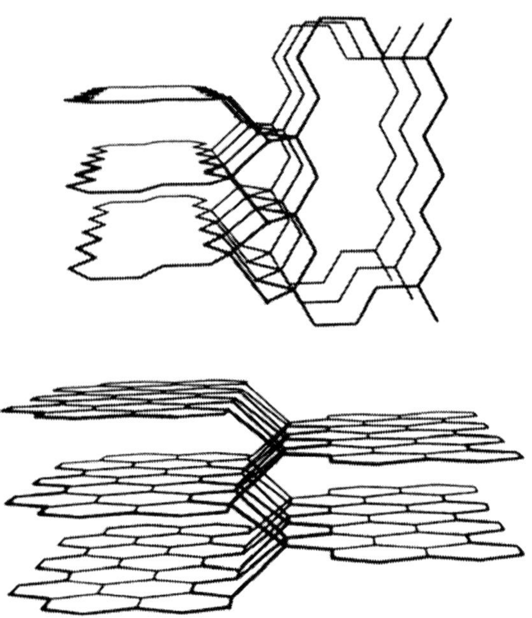

Figure 29. Representatives of the infinite families of ortho- and para- graphite-diamond hybrid structures, with the collective Wells point symbol (66)x(63)y, of orthorhombic symmetry (Pmmm).

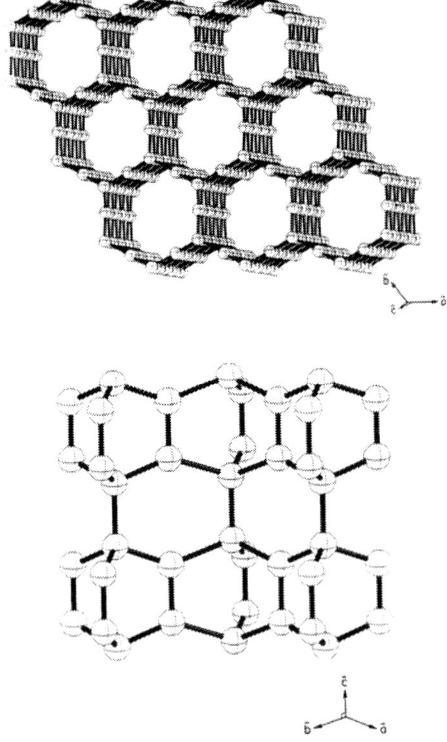

Figure 30. Vertical and lateral views of parent hexagonite structure of the infinite family of hexgonites, with the collective Wells point symbol (66)x(63)y, of orthorhombic-trigonal-hexagonal (Pmmm, P3m1 and P6/mmm) symmetries.

It should be noted here that the so-called "graphite-diamond hybrids", as described above, are models for novel types of allotropes of carbon. In the vein of this discussion, it should be mentioned here that the border-line space in the topology map of Figure 25, between the entry in the map of (5, 3), which is the pentagonal dodecahedron, and the entry in the map (6, 3) which is, of course, the graphene grid, lies the collective and infinite space of the fullerene C structures [131]. So, we can see that thus, the collective Schläfli symbol for the family of fullerenes is, in fact, given by $(5^{(x/(x+y))}, 3)$, where, in this instance, "x" is the number of pentagons in the polyhedron, and "y" is the number of hexagons in the polyhedron [117]. The Schläfli symbol for the parent C allotrope called "Buckminsterfullerene" (C_{60}) is $(5^{5/8}, 3)$ [131], and as substitution into Equation 6 will show, this Schläfli symbol rigorously describes the Buckminsterfullerene polyhedron fully. Figure 31 shows a view of this polyhedron of icosahedral symmetry, and it is clear from this view that the polyhedron is uniformly 3-connected, (as the Schläfli symbol reveals, the fullerenes are Archimedean) and comprised entirely of 5-gon and 6-gon circuitry.

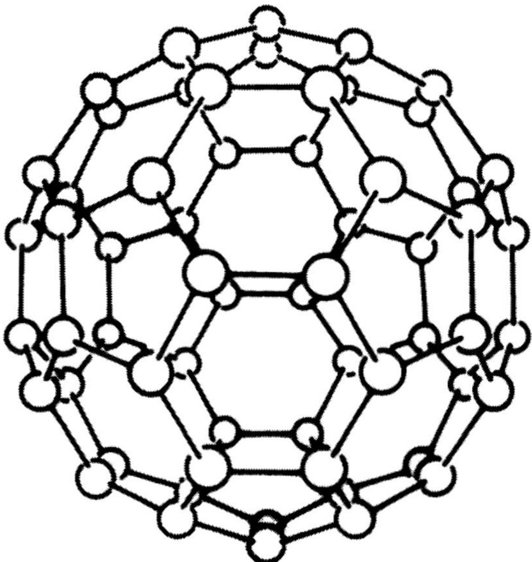

Figure 31. The parent Buckminsterfullerene polyhedron, of the infinite family of fullerenes, with the collective Schläfli symbol for the family of fullerenes given by (5(x/(x+y)), 3), and the fullerene polyhedral face symbol of 5x6y, where "x" is the number of hexagons, and "y" is the number of pentagons, in the fullerene, where such structures are of icosahedral (Ih) and lower symmetry.

Yet a final 3D network to be mentioned in this connection, which is a model of a 3-,4-connected network of C, as opposed to a straight "graphite-diamond" hybrid, lying between (6, 3) and (6, 4) of Figure 25, or a fullerene polyhedron, lying between (5, 3) and (6, 3) of Figure 25, as in the preceding discussions with respect to allotropes of C; is the so-called "glitter" network of C invented by Bucknum et al. in 1993-1994 [1, 132]. This structure can be envisioned as being constructed from a 1,4-cyclohexadiene building block, and it is shown in Figure 32.

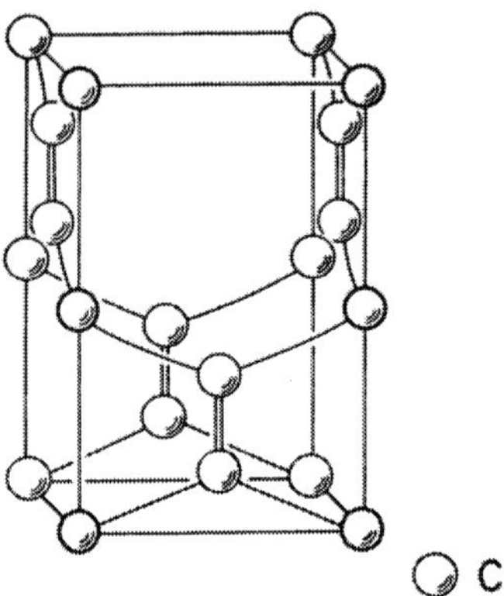

Figure 32. Tetragonal glitter network of carbon, with Wells point symbol given by (6284)(628)2, and of space group symmetry (P42/mmc).

As can be seen in Figure 32, the Wells point symbol for this Wellsean network is given by $(6^2 8^4)(6^2 8)_2$, and derived from this, is its Schläfli symbol of $(7, 3^{1/3})$ [116]. It is comprised of 6-gons admixed with 8-gons, in its topology, and an admixture of two trigonal vertices for every one tetrahedral vertex in its connection pattern. This particular C network has been important from the perspective of the 3-dimensional (3D) resonance structures which can be drawn over it, see Figure 33, and there have been some favorable indications that its synthesis has been achieved [133-134].

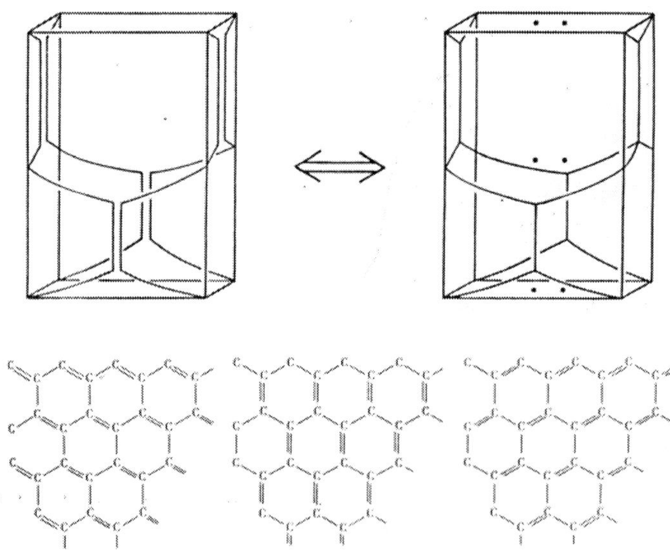

Figure 33. Resonance structures of the graphite and glitter networks of C.

Other inorganic networks that are of interest, that are not C allotropes, or models of C allotropes, would include the Archimedean Cooperite network, the structure of the minerals PtS and PdO [120], shown in Figure 34, which is a 4-connected network comprised of an equal mixture of tetrahedral and square planar vertices, both of which are distorted in their geometries [129].

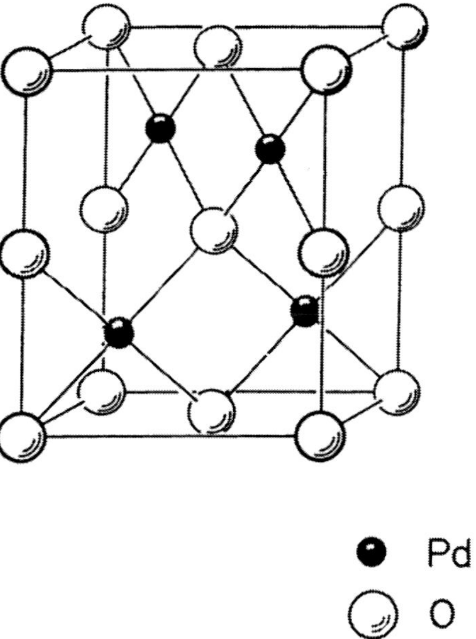

Figure 34. Archimedean Cooperite network as structure of PdO and PtS, with the Wells point symbol of (4284)(4282), and of space group symmetry (P42/mmc).

As Figure 34 indicates, taking both tetrahedral and square planar vertices as equally 4-connected, one can thus assign an Archimedean topology to this network, with a Wells point symbol of $(4^2 8^4)(4^2 8^2)$, and a Schläfli symbol of $(6^{2/5}, 4)$ [116]. In this case, the network nominally is binary, with two distinct types of connectivity, and an apparently Wellsean topology associated with this, but, in fact in this instance, we have a 4-connected network where the square planar vertices (with 4 independent circuits) are viewed as equivalent in topology (if distorted) versions of the tetrahedral vertices (with 6 independent vertices).

Yet another inorganic network includes, but is not limited to, the Catalan fluorite network [120], the structure of a number of mineral fluorides including CaF_2, shown in Figure 35.

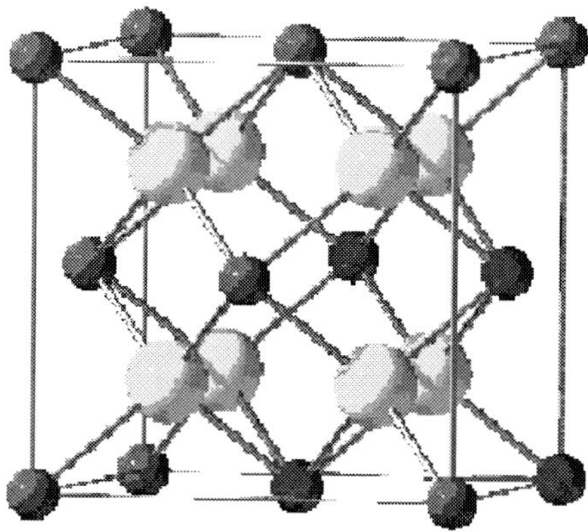

Figure 35. Catalan fluorite structure as the structure of CaF2, this network can be represented by the Wells point symbol (424)(46)2, and lies in space group (Fm-3m).

This densely connected network is comprised of 4-connected tetrahedral vertices, and 8-connected cube-centered vertices, which are connected to each other through a uniform set of 4-gons. The topology of this network can be represented by the Wells point symbol $(4^{24})(4^6)_2$, and this can be translated into a Catalan Schläfli symbol of $(4, 5^{1/3})$ [116]. It can be mapped in Figure 25 just beyond the entry (4, 5).

Still another inorganic structure-type we can provide the topology of, is the rocksalt (or primitive cubic) lattice [120], which is the structure of a number of inorganic alkali metal halides and alkaline earth chalcogenides. The rocksalt lattice is shown in Figure 36 below.

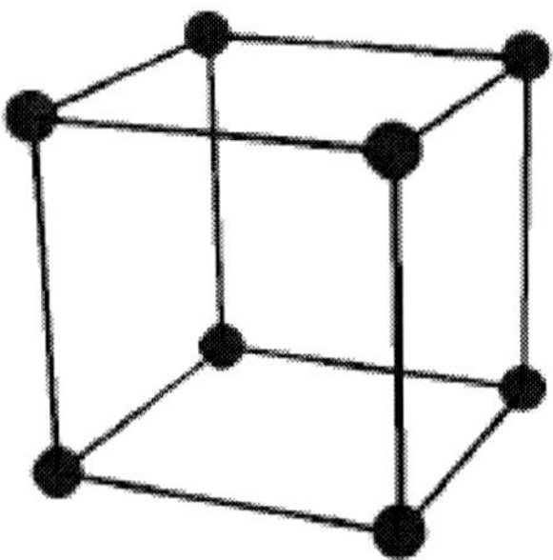

Figure 36. Platonic rocksalt structure as the structure of NaCl, this network can be represented by the Wells point symbol (412), and lies in space group (Pm-3m).

Also known as the primitive cubic structure-type, the rocksalt lattice is comprised uniformly of octahedral 6-connection in all 4-gon circuitry. The Wells point symbol for the network is given as 4^{12}, and this can be translated into a Schläfli symbol for the network of (4, 6) [116]. Because this rocksalt network is of a Platonic (regular, integer) topology, it can readily be seen where it maps in Figure 25. It represents, structurally, an extension into 3-dimensions (3D) of the square grid 4^4, or (4, 4) which extends in 2D, through a layering, in exact register, of other square grids onto a parent square grid, and their interconnection through perpendicular interlayer bonding through the respective vertices.

As a final inorganic structural-type that we can analyze here topologically in this Section, we have the so-called body-centered cubic (bcc) structure of CsCl [120], and a number of inorganic structures including alkali metal halides and alkaline earth chalcogenides and other materials. The bcc structure-type is shown here in Figure 37.

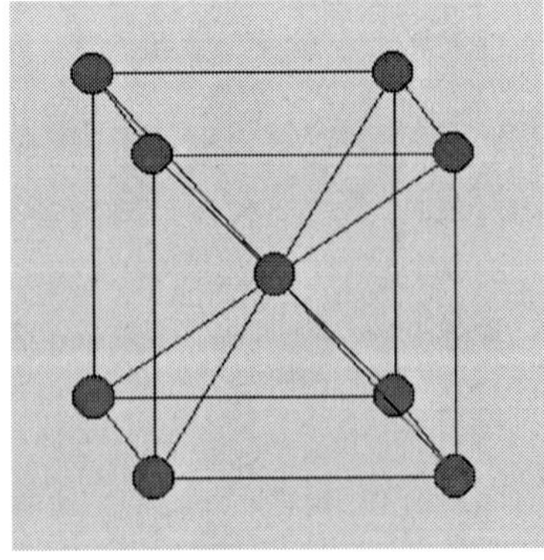

Figure 37. Platonic body-centered cubic (b.c.c.) structure as the structure of CsCl, this network can be represented by the Wells point symbol (424), and lies in space group (Im-3m).

It is uniformly comprised of 8-connected, cube-centered vertices that are mutually interconnected by all 4-gon circuits in the net. The topology of the bcc lattice can be specified by 4^{24}, or by the corresponding Platonic Schläfli symbol of (4, 8) [116].

There are, of course, innumerable other network structures that possess 3-dimensional geometries, one only has to look at the exhaustive works of O'Keeffe et al. to discern the scope of this field [135]. In the present discussion, we move, in the next Section, into the area of the topological analysis of some networks in 3-dimensions that exhibit rather odd Schläfli symbols (n, p). These networks, the oldest of which was identified only in 1951 (Waserite) [136], and the others of which were identified only in 1988 (Kentuckia) [137], and 2005 (Moravia) [138], suggest from the computation of their respective Schläfli symbols (n, p), by the methods outlined in Section 2 and Equation (3) above, that such numbers represent not only topological parameters of these networks, but they coincidently and fortuitously can be construed as rational approximations, in various instances, to the transcendental mathematical constants ϕ, e and π.

17. Rational Approximations to ϕ, E and π

Certain material networks, including the $CaCuO_2$ [137] (Kentuckia network [139]) structure-type, that is the progenitor for all of the superconducting cuprates, and the so-called Moravia structure-type [140], that is the proto-type structure for a number of coordination network (also called metal-organic frameworks or MOF's) structural compositions [138], and the so-called Waserite structure-type [136], that is the structure of the anionic, platinate sublattice of the ionic conducting lattice, known as sodium platinate ($NaPt_3O_4$), have here, through topological analysis, been shown, either to have a network polygonality, n, (Kentuckia network) the value of which serves as a rational approximation to the transcendental mathematical constant π, or, alternatively, they have a network connectivity, p, the value of which serves as a rational approximation to the product of the transcendental mathematical constants e and ϕ (Moravia network), or e and π (Waserite network).

The Kentuckia structure-type, proposed by Bucknum et al. in 2005 [139], is the pattern adopted by the high temperature superconducting cuprate composition, it is $CaCuO_2$ [137] that adopts this pattern, which is the progenitor to all the superconducting cuprates discovered so far. This tetragonal structure-type lies in space group P4/mmm, #123 and is shown in Figure 38 below.

It can be seen from Figure 38 that this tetragonal oxide, bears a relation to the cubic perovskite ($BaTiO_3$, [120]) structure-type which lies in the cubic space group Im-3m, not shown here, in which, by removal of an axial pair of oxygen vertices, one can generate the Kentuckia structure-type from the perovskite structure-type. However, it should be pointed out that the perovskite structure-type is a 6-,12-connected network, in which the transition metal titanium and chalcogenide oxygen centers, attain octahedral 6-coordination, while the alkaline earth barium cation bears a closest packed coordination sphere of 12. Whereas in the Kentuckia structure-type, the lattice, by great contrast as can be seen in Figure 38, bears an oxygen vertex with a square planar, 4-connected coordination, and the transition metal copper vertex bears an octahedral, 6-connected coordination, while the alkaline earth calcium cation is in cube-centered, 8-fold coordination. It is a ternary, 4-,6-,8-connected tetragonal structural pattern, as is described in Figure 38.

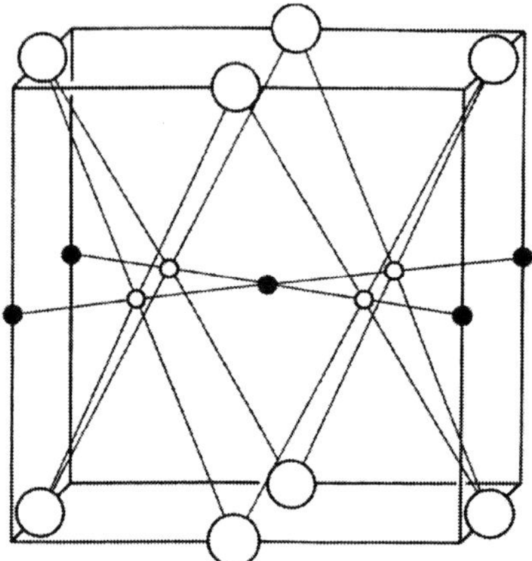

8-connected (body centered) B

6-connected (octahedral) C

4-connected (square planar) A

Figure 38. Wellsean Kentuckia (ABC2) structure-type as the structure of the superconducting cuprate salt CaCuO2, this network can be represented by the Wells point symbol (44)(412612)(412)2, and lies in space group (P4/mmm).

Thus the connectivity in Kentuckia, is that of a ternary 4-,6-,8-connected structural-type, while the connectivity in perovskite appears to be that of a binary 6-,12-connected network topology. It appears that all the circuitry in perovskite is comprised of 4-gons with, perhaps, the stoichiometry AB_4, in which the vertex “A” is in 12-connected, closest-packed topology, and the vertex “B” is 6-connected, octahedral coordination. This leads to a Wells point symbol for perovskite of $(4^{60})(4^{12})_4$ and a Schläfli symbol of (n, p) = $(4, 7^{1/5})$. Where, as a reference, the hexagonal closest packed (hcp) and cubic closest packed (ccp) networks, have the Wells point symbol 3^{60}, and a Schläfli symbol of (n, p) = (3, 12). In contrast, we see that the polygonality in the Kentuckia pattern, as revealed in Figure 38, is a composite of 4-gons and 6-gons, with the Wells point symbol for the lattice being $(4^4)(4^{12}6^{12})(4^{12})_2$. As Equation 10 reveals, the polygonality, n, for this structural-type, bears an odd resemblance to the transcendental mathematical constant π [142], occurring as it does, within 1% of exactly the value of $\sqrt{2}\cdot\pi$.

$$n = (40\cdot 4 + 12\cdot 6)/52 \qquad (10a)$$

$$n = \sqrt{2}\cdot\pi \qquad (10b)$$

Turning to the Waserite network [136], which is shown to be a relatively simple, binary 3-,4-connected network topology in Figure 39.

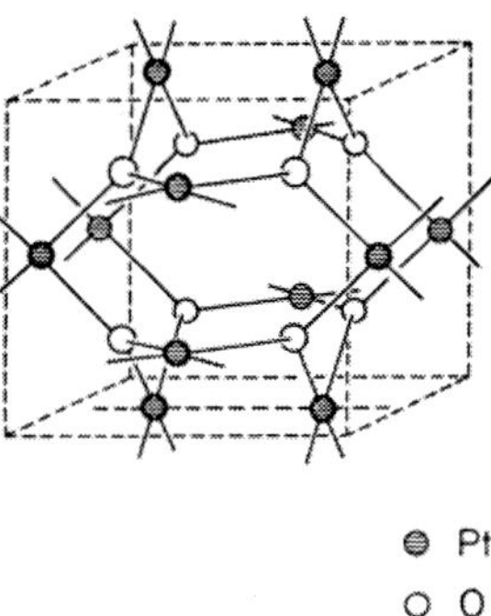

Figure 39. Catalan Waserite structure-type as the structure of the ionic conducting platinate salt NaPt3O4 (sodium cations not shown), this network can be represented by the Wells point symbol (84)3(83)4, and lies in space group (Pm-3n).

As is revealed in Figure 39, the 3-to-4 stoichiometry of 4-connected square planar vertices to 3-connected trigonal planar vertices, present in the Waserite topology, together with the Wells point symbol for this network as $(8^4)_3(8^3)_4$, thus demonstrates that this simple structure is indeed a binary, Catalan network comprised of all 8-gon circuitry. Therefore it is apparent, readily, that the polygonality is simply given by n = 8, in this pattern. But as has been described previously for this so-called Waserite network [141], the connectivity index of it, as shown in Equations 11, suggests that its topology is more complex than meets the eye.

$$p = (3\cdot4 + 4\cdot3)/7 \quad (11a)$$

$$p = (2/5)e\cdot\pi \quad (11b)$$

As Equation 11a-b reveals, it is a fact of simple arithmetic that the weighted average connectivity of the Waserite network, given by the symbol p, is in fact equal, to better than 99 parts in 100, to (2/5)e·π. Here π is, the familiar ratio of a circle's circumference to its diameter [142], and e is the natural base of exponentials [143]. These numbers are transcendental, as they are infinite, non-repeating fractions [142, 143]. An identical relation will also hold for the other structures patterned on a stoichiometry of four 3-connected vertices-to-three 4-connected vertices including, for example, the rhombic dodecahedron (given by the Catalan Wells point symbol as $(4^4)_6(4^3)_8$), and the well-known phenacite network of Bragg et al. (not shown, given by the Wellsean ternary Wells point symbol $(8^3)(6^3)_3(6^38^3)_3$ [144])

Other relations emerging from such consideration of the connectivity index, p, in the Waserite structural-type include Equation 12 [141], known hereafter as the Timaeus relation, for its suggestion of a Cosmogony based upon the 5 Platonic solids as enunciated by Plato [118].

$$(1)\cdot(2.3333333............)\cdot e\cdot\pi = (4)\cdot(5) \quad (12)$$

Equation (12) suggests the ultimate simplicity of definitions of e and π, through an elementary relationship involving only the first 5 counting numbers, or alternatively, the first 4 prime numbers [141].

Finally, in this survey of crystalline structure-types which exhibit relations to the transcendental mathematical constants, in their structural topology, we turn to the so-called Moravia network [138, 140], first posited as a potential structural-type in 2005 by Bucknum et al. This Moravia network has, in fact, turned out to be the structure adopted by several coordination networks known as metal-organic-frameworks (MOF's) [138]. It is readily seen to be a Wellsean, 3-,8-connected network upon careful inspection of the drawing for valences, and tracing of circuitry in Figure 40.

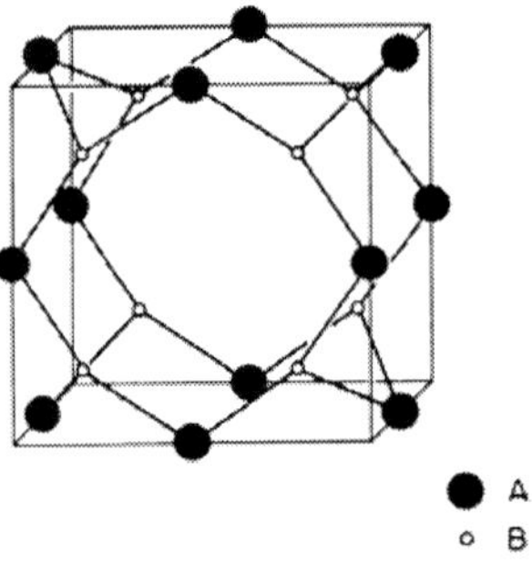

Figure 40. Wellsean Moravia structure-type as the structure of several coordination networks (metal organic frameworks, MOF's), this network can be represented by the Wells point symbol (4468812)3(43)8, and lies in space group (Pm-3m).

The Wells point symbol for the Moravia structural-type is encoded as $(4^4 6^8 8^{12})_3(4^3)_8$, it is thus a complex, Wellsean network composed of two connection motifs, the trigonal planar, 3-connected, and cube-centered, 8-connected, vertices, held together in circuits of 4-gon, 6-gon and 8-gon sizes. The complex Well's point symbol for the network, belies in this instance, the relatively high symmetry of the structure, in which Moravia is lying in the cubic space group Pm-3m, #221.

If we take the point symbol for the network, and analyze it according to Equations 7a-b described above, the Well's point symbol translation formulas, one obtains the result that the weighted average polygonality for the network is indeed given by n = 6. The Waserite net is thus pseudo-Catalan, with an integer polygonality of 6, which is nonetheless the result of averaging over 4-, 6- and 8-gons in its structural pattern. Upon calculating the connectivity index, p, for Moravia, however, we get the result shown in Equation 13a-b.

$$p = (3 \cdot 8 + 8 \cdot 3)/11 \tag{13a}$$

$$p = e \cdot \phi \tag{13b}$$

Thus it is seen that the connectivity index, p, of Moravia, as a 3-,8-connected network, is equivalent, to better than 99 parts in 100, to the product of the two transcendental mathematical constants ϕ [145] and e [143], given simply as $\phi \cdot e$. Here, as above, e is the natural base of exponentials [143], and ϕ is the well-known golden ratio [145], as is expressed in Equation 14.

$$\phi = (\sqrt{5} + 1)/2 \tag{14}$$

Equations 13a-b, like the transformation of Equations 11a-b to Equation 12 above, can be factored, interestingly, so that the relation of $\phi \cdot e$ evolving out of the topology of the Moravia structure, as shown in Figure 40 above, involves the first 6 Fibonacci numbers, F(1-to-6) (given, on the left, in Equation 15b as 1, 1, 2, 3, 5, and 8), and these are related to the 10th Fibonacci number, F(10) (given, on the right, in Equation 15b as 55).

$$F(1)\cdot F(2)\cdot F(3)\cdot F(4)\cdot F(5)\cdot F(6) = (e)\cdot(\phi)\cdot F(10) \qquad (15a)$$

$$(1)\cdot(1)\cdot(2)\cdot(3)\cdot(5)\cdot(8) = (e)\cdot(\phi)\cdot(55) \qquad (15b)$$

These relations between the topology of these structures in this Section, as is revealed by the computation and mapping of their corresponding Schläfli symbols (n, p), and the transcendental mathematical constants ϕ, e and π, that can be thus correlated to their structural character, suggest the mathematical, and potentially scientific, richness that such structures may lead to.

18. Conclusions

In this paper we have reviewed the basic tenets of a chemical topology scheme, one that can be applied to classify and effectively map the innumerable polyhedra, tessellations and networks, based upon a simple computation of their Schläfli symbols (n, p), from translation of their corresponding Wells point symbols. A restriction pointed up by this work, is that all structures in such a chemical topology scheme must, indeed, be *simply connected.* The phrase *"simply connected"* means that all edges, E, in a structure, be it a polyhedron, tessellation or network, must terminate at distinct vertices, V, in the network, where such edges, E, are known as proper edges. A second condition on a net being *"simply connected"*, is that all faces, F, in the structure should be bounded by proper edges, E, as defined in the preceding sentence. It is not clear, at this juncture, what form a systematic chemical topology would take on for the *"non-simply connected"* structures. As there are innumerable *non-simply connected* structures, to accompany the infinite number of *simply connected* structures in Schläfli space (the space of the Schläfli symbols (n, p)), it would seem that a topological analysis of these complex structures would be desirable and necessary to get a more complete handling of the chemical topology of crystal chemistry.

The brunt of this paper has been dedicated to a survey of some of the more prominent (well-known or obvious) organic and inorganic structure-types. Organic structures included some well-known C allotropes, like the regular, graphene grid and the regular, diamond network, both forms of C known since Antiquity. And, also, more modern C forms were surveyed, like the 3-,4-connected, Catalan graphite-diamond hybrids [128], the 3-,4-connected, Catalan hexagonite lattices [129, 130] and the Wellsean, 3-,4-connected glitter C form [132], for which there is currently some evidences of their syntheses from the growth of C nanocrystals [133, 134]. Inorganic structures included in this survey, were the 4-,8-connected, Catalan fluorite lattice, the 4-connected, Archimedean Cooperite lattice, the 8-connected, regular CsCl, body-centered cubic structure-type, and the 6-connected, regular rocksalt structure-type. [120] Finally, in this survey, lattices which admitted connections in their topology to the transcendental numbers included the 3-,8-connected, Wellsean Moravia net, discovered in 2005 [138, 140] (related to ϕ and e, through the connectivity), the 4-,6-,8-connected, Wellsean Kentuckia (cuprate structure-type) net, discovered in 1988 [137, 139] (related to π, through the polygonality), and finally the 3-,4-connected, Catalan Waserite net (platinate structure-type), discovered in 1951 [136].

The occurrence of relations to the transcendental numbers of mathematics, in the computations of the topology character of some of these networks, is indeed a mysterious outcome. It is not clear whether such relations could imply that the topology of these lattices, like the Kentuckia lattice, in which $n = \sqrt{2}\cdot\pi$, could indeed be equated to some type of ordering parameter for the lattice, such that by the introduction of systematic defects in the connectivity, p (or thereby the polygonality, n) over the bulk lattice, might lead to a corrected value of n, that asymptotically approaches the true value of π, and that, that might have some bearing on bulk properties of the Kentuckia network, like the critical superconducting transition temperature, T_c, in the cuprate composition $CaCuO_2$ [139]. Such considerations as these, open up new avenues of explorations for solid state scientists based upon the intrinsic topology character of such networks as these.

REFERENCES (PART I)

[1] (a) United States Patent & Trademark Office, M.J. Bucknum, Disclosure Document #337,466, p. 1-20, August 18th, 1993, and (b) M.J. Bucknum and R. Hoffmann, *Journal of the American Chemical Society*, 116, 11456, (1994).

[2] (a) L.A. Carreira, R.O. Carter and J.R. Durig, Journal of Chemical Physics, 59, 812, (1973) (b) H. Oberhammer and S.H. Bauer, Journal of the American Chemical Society, 91, 10, (1969). *The structural information is based upon electron diffraction data of the gaseous 1,4-cyclohexadiene molecule. Carreira et al. report the ring in a planar equilibrium conformation.*

[3] A.F. Wells, The Geometrical Basis of Crystal Chemistry: (a) Part 1, A.F. Wells, *Acta Cryst.*, 7, 535, (1954). (b) Part 2, A.F. Wells, *Acta Cryst.*, 7, 545, (1954). (c) Part 3, A.F. Wells, *Acta Cryst.*, 7, 842, (1954). (d) Part 4, A.F. Wells, *Acta Cryst.*, 7, 849, (1954). (e) Part 5, A.F. Wells, *Acta Cryst.*, 8, 32, (1955). (f) Part 6, A.F. Wells, *Acta Cryst.*, 9, 23, (1956). (g) Part 7, A.F. Wells and R.R. Sharpe, *Acta Cryst.*, 16, 857, (1963). (h) Part 8, A.F. Wells, *Acta Cryst.*, 18, 894, (1965). (i) Part 9, A.F. Wells, *Acta Cryst.*, B24, 50, (1968). (j) Part 10, A.F. Wells, *Acta Cryst.*, B25, 1711, (1969). (k) Part 11, A.F. Wells, *Acta Cryst.*, B28, 711, (1972). (l) Part 12, A.F. Wells, *Acta Cryst.*, B32, 2619, (1976). (m) A.F. Wells, Three Dimensional Nets and Polyhedra, 1st edition, Wiley and Sons, New York, 1977. (n) A.F. Wells, Further Studies of Three-dimensional Nets, American Crystallographic Association, Monograph #8, 1st edition, ACA Press, 1979.

[4] K.M. Merz, R. Hoffmann and A.T. Balaban, *Journal of the American Chemical Society*, 109, 6742, (1987).

[5] (a) R. Hoffmann, *Journal of Chemical Physics*, 39, 1397, (1963). (b) R. Hoffmann and W.N. Lipscomb, *Journal of Chemical Physics*, 37, 2872, (1962). (c) M-H. Whangbo and R. Hoffmann, *Journal of the American Chemical Society*, 100, 6093, (1978). (d) M-

H. Whangbo, R, Hoffmann and R.B. Woodward, *Proceedings of the Royal Society*, A366, 23, (1979).

[6] (a) H.E. Simmons and T. Fukunaga, *Journal of the American Chemical Society*, 89, 5208, (1967). (b) R. Hoffmann, A. Imamura and G. Zeiss, *Journal of the American Chemical Society*, 89, 5215, (1967).

[7] (a) V. Galasso, *Chemical Physics*, 153, 13, (1991). (b) S. Smolinski, M. Balazy, H. Iwamura, T. Sugawara, Y. Kawada and M. Iwamura, *Bulletin of the Chemical Society of Japan*, 55, 1106, (1982). (c) H. Duerr and R. Gleiter, *Angewandte Chemie International Edition in English*, 17, 559, (1978). (d) B.Y. Simkin, S.P. Makarov, N.G. Furmanova, K.S. Karaev and V.I. Minkin, *Chemistry of Heterocyclic Compounds*, 1, 948, (1978). (e) J.S. Foos, "Synthesis and Reactivity of Spiro[4.4]nonatetraene and Spiro[4.4]nona-1,3,6-triene", PhD thesis, Cornell University, 1974. (f) C. Batich, E. Hielbronner, E. Rommel, M.F. Semmelhack and J.S. Foos, *Journal of the American Chemical Society*, 96, 7662, (1974). (g) R. Sustmann and R. Schubert, *Angewandte Chemie International Edition in English*, 11, 840, (1972). (h) H. Duerr and H. Kober, *Justus Liebigs Ann. Chem.*, 740, 74, (1970). (i) E.W. Garbisch, Jr. and R.F. Sprecher, *Journal of the American Chemical Society*, 88, 3433, (1966). (j) P.E. Eaton and R.A. Hudson, *Journal of the American Chemical Society*, 87, 2769, (1965). (k) K.N. Houk, *Journal of the American Chemical Society*, 87, 2769, (1965).

[8] V.I. Minkin, M.N. Glukhovtsev and B.Y. Simkin, *Aromaticity and Antiaromaticity: Electronic and Structural Aspects,* Wiley and Sons, NY, 1994.

[9] A.F. Wells, *Structural Inorganic Chemistry*, Oxford University Press, Oxford, UK, 1984.

[10] J.K. Williams and R.E. Benson, *Journal of the American Chemical Society*, 84, 1257, (1962).

[11] R.E. Benson and R.V. Lindsey, Jr., *Journal of the American Chemical Society*, 81, 4247, (1959).

[12] R.B. Woodward and R. Hoffmann, *The Conservation of Orbital Symmetry,* Verlag Chemie, GmbH, Weinheim/Bergstrasse, 1970.

[13] (a) J. Donohue, The Structure of the Elements, Wiley and Sons, NY, 1974. (b) W. Kraetschmer and H. Huffmann, *Nature*, 347, 354, (1990). (c) H. Kroto, J.R. Heath, S.C. O'Brien, R.F. Curl and R.E. Smalley, *Nature*, 318, 162, (1985). (d) E. Osawa, *Kagaku*, 25, 85, (1970). (e) F.P. Bundy and J.S. Kasper, *Journal of Chemical Physics*, 46, 3437, (1967). (f) R.B. Aust and H.G. Drickamer, *Science*, 140, 817, (1963). (g) L. Zeger and E.Kaxiras, *Physical Review Letters*, 70, 2929, (1993). (h) D.A. Muller, Y. Zhou, R. Raj, and J. Silcox, *Nature*, 366, 725, (1993). (i) H. Hiura, T.W. Ebbensen, J. Fujita, K. Tanigaki and T. Takada, *Nature*, 367, 148, (1994). (j) C.L. Renschler, J. Pouch and D.M. Cox, editors, Novel Forms of Carbon, MRS Symposium Proceedings, 270, MRS Pittsburgh, 1992. (k) F. Moshary, N.H. Chen, I.F. Silvera, C.A. Brown, H.C. Dorn, M.S. de Vries and D.S. Bethune, *Physical Review Letters*, 69, 466, (1992). (l) S. Iijima, *Nature*, 354, 56, (1991). (m) W. Utsumi and T. Yagi, *Science*, 252, 1542, (1991). (n) H. Hirai and K. Kondo, *Science*, 253, 772, (1991). (o) K.E. Spear, A.W. Phelps and W.B. White, *Journal of Materials Research*, 5, 2277, (1990). (p) A.V. Baitin, A.A. Lebedev, S.V. Romanenko, V.N. Senchenko and M.A. Scheindlin, *High Temperatures-High Pressures*, 21, 157, (1990). (q) J.C. Angus and C.C. Hayman, *Science*, 241, 913, (1988). (r) V.M. Melnitchenko, Y.N. Nikulin and A.M. Sladkov, *Carbon*, 23, 3, (1985).

(s) I.V. Stankevich, M.V. Nikerov and D.A. Bochvar, *Russian Chemical Reviews*, 53, 640, (1984). (t) A.G. Whittaker, E.J. Watts, R.S. Lewis and E. Anders, *Science*, 209, 1512, (1980). (u) A.G. Whittaker and B. Tooper, *Journal of the American Ceramic Society*, 57, 443, (1974). (v) D.A. Bochvar and E.G. Galpern, *Dokl. Akad. Nauk. SSSR*, 209, 612, (1973). (w) A. El Gorsey and G. Donnay, *Science*, 161, 363, (1968). (x) S. Ergun, *Carbon*, 6, 141, (1968). (y) H. Drickamer, *Science*, 156, 1183, (1967). (z) F.P. Bundy, *Journal of Chemical Physics*, 38, 631, (1963).

[14] R.M. Hazen, *The New Alchemists*, Times Books-Random House, New York, 1994.

[15] P.W. Atkins, *Physical Chemistry*, 4th edition, W.H. Freeman and Company, New York, 1990.

[16] *More descriptive topological indexes are sometimes used in place of Schlaefli symbols because of the need to further identify nets with the same Schlaefli symbol (n, p).* For example the graphite net has the index (6, 3), which is similar to the index for the square-octagon plane tiling and the pentagon-heptagon plane tiling.

[17] *Elementa doctrinae solidorum and Demonstratium nonannularum in signium proprietatum quibus solid heddris planis inclusas unt praedita*, L. Euler, included in the Proceedings of the St. Petersburg Academy, 1758.

[18] *Explicit Euler equations for 3-, 4- and 5-connected polyhedra elegantly show the existence of the 5 regular convex polyhedra and the semi-regular* Archimedean polyhedra.

[19] *For 1-dimension, n and p can be viewed as each being equal to 2. (2, 2) has a number of edges, E, equal to 2. For n-sided polygons, the Schlaefli symbol is (n, 2) and the number of edges, E, is properly equal to n.*

[20] A.F. Wells, *The Geometrical Basis of Crystal Chemistry: Part 7, A.F. Wells and R.R. Sharpe,* Acta Cryst., 16, 857, (1963). This paper reports on a modification of the Euler relation for application to 3-dimensional polyhedra. Such polyhedra are infinite in extent and are represented by the Schlaefli symbols (3, n) where n is greater than 6, and other Schlaefli symbols, some of which are identical to those of the ordinary 3-dimensional nets.

[21] M. Wenninger, *Dual Models*, 1st edition, Cambridge University Press, Cambridge, UK, 1983. And the references therein.

[22] A.F. Wells, *Further Studies of Three-dimensional Nets, ACA Monograph #8,* 1st edition, ACA Press, 1979. And the references therein. Much previously unpublished material, concerned mainly with the vast topological subspace of 3-,4-connected networks, was reported in this publication. Scores of previously unknown structures, perhaps numbering more than 100, were reported. The explicit identification of structures with both fractional polygonality and fractional connectivity was made, anticipating much later work in various fields. These Wellsean structures are innumerable and their polyhedral analogs have recently been identified.

[23] W.L. Bragg and W.H. Zachariasen, *Zeitschrift fuer Kristolograffie*, 72, 518, (1930).

[24] W.L. Bragg, *Atomic Structure of Minerals*, Cornell University Press, Ithaca, NY, 1937.

[25] (a) R. Juza and H. Hahn, *Naturwissenschaften*, 27, 32, (1939). (b) R. Juza and H. Hahn, *Zeitschrift fuer Anorganische Chemie*, 244, 125, (1940).

[26] D. Hardie and K.H. Jack, *Nature*, 180, 332, (1957).

[27] S.N. Ruddleson and P. Popper, *Acta Cryst.*, 11, 465, (1958).

[28] M.L. Cohen, *Physical Review B*, 32, 7988, (1985).

[29] J. Waser and E.D. McClanahan, *Journal of Chemical Physics*, 19, 413, (1951).

References (Part II)

[30] I. Konyashin, A. Zern, J. Mayer, *Diamond and Related Materials*, 10, 99, (2001).

[31] L.S. Palatnik, M.B. Guseva, V.G. Babaev, N.F. Savchenko, L.L. Fal'ko, *Soviet Physics, JETP*, 60, 520, (1984).

[32] Y. Gogotsi, S. Weiz, D.A. Ersoy and M.J. McNallan, *Nature (London)*, 411, 283, (2001).

[33] H. Hirai and K. Kondo, *Science*, 253, 772, (1991).

[34] M.J. Bucknum and R. Hoffmann, *Journal of the American Chemical Society*, 116, 11456, (1994).

[35] M.J. Bucknum, *Carbon*, 35, 1, (1997).

[36] M.J. Bucknum and E.A. Castro, *Journal of Mathematical Chemistry*, 36(4), 381, (2004).

[37] L. Bursill et al., *International Journal of Modern Physics B*, 15, 3107, (2001).

[38] J.E. McMurry, *Organic Chemistry*, 6th edition, Brooks-Cole, New York, 2004.

[39] *Although as is reported here and elsewhere, a hypothetical metallic modification of carbon with a density of about 3.1-3.2 g/cm3, called glitter,* has been reported by Bucknum et al. in 1994. See references [5-7].

[40] C.J. Pickard, V. Milman and B. Winkler, *Diamond and Related Materials*, 10, 2225, (2001).

[41] P. Hohnenberg and W. Kohn, *Physical Review B*, 136, 864, (1964).

[42] R. Hoffmann, *Journal of Chemical Physics*, 39, 1397, (1963).

[43] R. Hoffmann and W.N. Lipscomb, *Journal of Chemical Physics*, 37, 2872, (1962).

[44] M.H. Whangbo and R. Hoffmann, *Journal of the American Chemical Society*, 100, 6093, (1978).

[45] M.H. Whangbo, R. Hoffmann and R.B. Woodward, *Proceedings of the Royal Society A*, 366, 23, (1979).

[46] CASTEP (*Cambridge Serial Total Energy Package) was created and developed by Professor M.C. Payne and others in the 1980's. It is a computational procedure based upon the density functional theory (DFT)* referenced in [12]. Professor M.C. Payne and his colleagues are located in the Theoretical Condensed Matter group in the Cavendish laboratory at Cambridge University. Dr. Chris J. Pickard kindly performed and provided the calculations on the glitter structure to the authors. The details of the CASTEP program are described in *Rev. Mod. Phys.*, 64, 1045, (1992). For the present implementation of CASTEP, used to optimize the structural parameters of glitter, ultrasoft pseudopotentials were employed, the basis set had an energy cutoff of 400 eV and k-point sampling was done with a 10x10x4 mesh.

[47] L.A. Carreira, R.O. Carter and J.R. Durig, *Journal of Chemical Physics*, 59, 812, (1973).

[48] H. Oberhammer and S.H. Bauer, *Journal of the American Chemical Society*, 91, 10, (1969).

[49] G. Dallinga and L.H. Toneman, *Journal of Molecular Structure*, 1, 117, (1967).

[50] A.F. Wells, *Three Dimensional Nets and Polyhedra*, 1st edition, John Wiley and Sons, New York, 1977.

[51] A.F. Wells, *Further Studies of Three-dimensional Nets, ACA Monograph #8,* American Crystallographic Association Press, Pittsburgh, PA, 1979.

[52] R. Hoffmann, T. Hughbanks, M. Kertesz and P.H. Bird, *J. Am. Chem. Soc.*, 105, 4831, (1983).

[53] L. Pauling, *The Nature of the Chemical Bond*, 3rd edition, Cornell University Press, Ithaca, New York, 1960.

[54] *Theoretical diffraction data for the glitter lattice, including intensity calculations, with lattice parameters for the tetragonal cell of a = 2.530 Å and c = 5.980* Å, have been reported previously. See M.J. Bucknum, Chemistry Preprint Archive, 2000, 11, 10-62, (2000), and, M.J. Bucknum and E.A. Castro, Journal of Molecular Modeling, in press, (2005).

[55] *The Amsterdam Density Functional program (ADF2002.02) was employed in the calculations of the glitter band structure and DOS.* Details of this program package can be found in the following reference, G. te Velde, F.M. Bickelhaupt, E.J. Baerends, C.F. Guerra, S.J.A. van Gisbergen, J.G. Snijders and T. Ziegler, Journal of Computational Chemistry, 22, 931, (2001).

[56] *These were thoughts made in a personal communication with the one of the authors (M.J.B.)* by Professor Howard Simmons at Du Pont Central Research & Development (CRD).

[57] H.E. Simmons and T. Fukunaga, *J. Am. Chem. Soc.,* 89, 5208, (1967).

[58] R. Hoffmann, A. Imamura and G. Zeiss, *J. Am. Chem. Soc.*, 89, 5215, (1967).

[59] P. Maslak, *Advanced Materials*, 6(5), 405, (1994).

[60] V. Galasso, *Chem. Phys.*, 153, 13, (1991).

[61] S. Smolinski, M. Balazy, H. Iwamura, T. Sugawara, Y. Kawada and M. Iwamura, *Bull. Chem. Soc. of Japan,* 55, 1106, (1982).

[62] H. Dürr and R. Gleiter, *Angew. Chem. Int. Ed. Engl.*, 17, 559, (1978).

[63] B.Y. Simkin, S.P. Makarov, N.G. Furmanova, K.S. Karaev and V.I. Minkin, *Chem. Heterocycl. Compds.*, 948, 1978.

[64] J.S. Foos, S*ynthesis and Reactivity of Spiro[4.4]Nonatetraene and Spiro[4.4]Nona-1,3,6-Triene,* PhD thesis, Cornell University, 1974.

[65] C. Batich, E. Heilbronner, E. Rommel, M.F. Semmelhack and J.S. Foos, *J. Am. Chem. Soc.*, 96, 7662, (1974).

[66] R. Sustmann and R. Schubert, *Angew. Chem. Internat. Edit. Engl.,* 11, 840, (1972).

[67] H. Dürr and H. Kober, *Justus Liebigs Ann. Chem.*, 740, 74, (1970).

[68] E.W. Garbisch, Jr. and R.F. Sprecher, *J. Am. Chem. Soc.*, 88, 3433, (1966).

[69] P.E. Eaton and R.A. Hudson, *J. Am. Chem. Soc.,* 87, 2769, (1965).

[70] K.N. Houk, *J. Am. Chem. Soc.*, 87, 2769, (1965).

[71] E. Osawa, *Kagaku*, 25, 85, (1970).

[72] D.E.H. Jones, *New Scientist*, 32, 245, (1966).

[73] H.W. Kroto, J.R. Heath, S.C. O'Brien, R.F. Curl and R.E. Smalley, *Nature(London)*, 318, 162, (1985).

[74] M.J. Bucknum, *Chemistry Preprint Archive*, 2001(1), 28-63, (2001).

[75] A.T. Balaban, D.J. Klein and C.A. Folden, *Chem. Phys. Lett.*, 217, 266, (1994).

[76] R.H. Baughman and D.S. Galvao, *Chem. Phys. Lett.*, 211, 110, (1993)

[77] R.H. Baughman and D.S. Galvao, *Nature(London)*, 365, 735, (1993).
[78] K. Tanaka, K. Okahara, M. Okada and T. Yamabe, *Chem. Phys. Lett.*, 191, 469, (1992).
[79] H.R. Karfunkel and R. Dressler, *J. Am. Chem. Soc.*, 114, 2285, (1992).
[80] F. Diederich and Y. Rubin, *Angew. Chem., Int. Ed. Engl.*, 31, 1101, (1992).
[81] D. Boercker, *Phys. Rev. B*, 44, 11592, (1991).
[82] A.Y. Liu, M.L. Cohen, K.C. Hass and M.A. Tamor, *Phys. Rev. B*, 43, 6742, (1991).
[83] C. Mailhot and A.K. McMahan, *Phys. Rev. B*, 44, 11578, (1991).
[84] G. Laqua, H. Musso, W. Boland and R. Ahlrichs, *J. Am. Chem. Soc.*, 112, 7391, (1990).
[85] M.A. Tamor and K.C. Hass, *J. Mater. Res.*, 5, 2273, (1990).
[86] R.L. Johnston and R. Hoffmann, *J. Am. Chem. Soc.*, 111, 810, (1989).
[87] R.H. Baughman, H. Eckhardt and M. Kertesz, *J. Chem. Phys.*, 87, 6687, (1987).
[88] J. Robertson and E.P. O'Reilly, *Phys. Rev. B*, 35, 2946, (1987).
[89] K.M. Merz, R. Hoffmann and A.T. Balaban, *J. Am. Chem. Soc.*, 109, 6742, (1987)
[90] A.T. Balaban, *Comput. Math. Appl.*, 17, 397, (1987).
[91] S.E. Stein and R.L. Brown, *J. Am. Chem. Soc.*, 109, 3721, (1987).
[92] J.K. Burdett and S. Lee, *J. Am. Chem. Soc.*, 107, 3050, 3063, 3083, (1985).
[93] R. Biswas, R.M. Martin, R.J. Needs and O.H. Neilsen, *Phys. Rev. B*, 30, 3210, (1984).
[94] R. Hoffmann, T. Hughbanks, M. Kertesz and P.H. Bird, *J. Am. Chem. Soc.*, 105, 4831, (1983).
[95] M. Kertesz and R. Hoffmann, *J. Solid State Chem.*, 54, 313, (1980).
[96] R.A. Davidson, *Theor. Chim. Acta*, 58, 193, (1981).
[97] R. Hoffmann, O. Eisenstein and A.T. Balaban, *Proc. Natl. Acad. Sci. U.S.A.*, 77, 5588, (1980).
[98] A.T. Balaban, C.C. Rentia and E. Ciupitu, *Rev. Roum. Chim.*, 231, 1233, (1968).
[99] J. Kakinoki, *Acta Crystallogr.*, 18, 578, (1965).
[100] H.L. Riley, *J. Chim. Phys.*, 47, 565, (1950).
[101] D.E. Sands, *Introduction to Crystallography*, 1st edition, Dover Science Books, New York, 1993.
[102] M.J. Bucknum, *Chemistry Preprint Archive*, 2000(11), 10-62, (2000).
[103] M.J. Bucknum, *Chemistry Preprint Archive*, 2004(4), 54-79, (2004).
[104] M.J. Bucknum and E.A. Castro, *Journal of Molecular Modeling*, 12, 111, (2005).
[105] M.J. Bucknum and E.A. Castro, *Journal of Mathematical Chemistry*, in press, (2005).
[106] R.M. Hazen, The New Alchemists, Times Books-Random House, New York, 1994.
[107] I. Stamatin, A. Dumitru, M.J. Bucknum, V. Ciupina and G. Prodan, *Molecular Crystals & Liquid Crystals*, 417, 167, (2004).

REFERENCES (PART III)

[108] M.J. Bucknum, *Carbon*, 35(1), 1, (1997).
[109] Leonhard Euler, *Elementa doctrinae solidorum et Demonstratio nonnularum insignium proprietatum quibus solida heddris planis inclusa sunt praedita,* Proceedings of the St. Petersburg Academy, St. Petersburg, Russia 1758.
[110] J.J. Burckhardt, *Der mathematische Nachlass von Ludwig Schläfli, 1814-1895, in der Schweizerischen Landesbibliothek*, 1st edition, Bern, 1942.

[111] A.F. Wells, *Three Dimensional Nets and Polyhedra*, 1st Edition, John Wiley and Sons Inc., New York, NY 1977.

[112] M. Gardner, *Archimedes: Mathematician and Inventor*, 1st edition, Macmillan Publishing, New York, NY, 1965.

[113] P.J. Frederico, *Descartes on Polyhedra: A Study of the De Solidorum Elementis*, 1st edition, Springer-Verlag, New York, NY, 1982.

[114] A.F. Wells, *Further Studies of Three-dimensional Nets, American Crystallographic Association (A.C.A), Monograph* #8, 1st Edition, A.C.A Press, Pittsburgh, PA 1979.

[115] I. Peters, *Science News*, 160(25/26), 396, (2001).

[116] M.J. Bucknum and E.A. Castro, *(MATCH) Commun. Math. Comp. Chem.*, 54, 89, (2005).

[117] It appears that the 5 Platonic polyhedra obey Equation (2) of the text, if one specifies their topology by a Well's point symbol, given by A^a, in which n = A and p = a, and there is no relation between this polyhedral face symbol, A^a, and the computation of V, E and F in the Euler model of Equation (1). It is also the case, that the 5 Platonic polyhedra obey Equation (2) of the text, if one specifies their topology by a polyhedral face symbol, given as A^b, in which (b·A) = 2E, b = F, and V is identified through inspection of the polyhedron, then n = 2E/F and p = 2E/V, by definition. For the Archimedean polyhedra it appears that it is only possible to specify (n, p), for insertion into Equation (2), by encoding a polyhedral face symbol, given as $A^aB^bC^c$......., where a is the number of A-gons, in a ratio with b B-gons etc., for the polyhedron, and thus, in which (a·A + b·B + c·C + ...) = 2E, (a + b + c + ...) = F, and V is identified through inspection of the polyhedron, and where finally then, n = 2E/F and p = 2E/V, by definition. While for the Catalan and Wellsean polyhedra, by contrast, it appears that it is only possible to specify (n, p), for insertion into Equation (2), by encoding a Wells point symbol, given as $(A^a)_x(A^b)_y(A^c)_z$.........., where a is the number of A-gons meeting at a, and b is the number of A-gons meeting at b etc., and x, y, z etc. specify the stoichiometry of the polyhedron, for the Catalan or Wellsean polyhedron of interest, in which (a·A + b·A + c·A + ...) = E, (a·x + b·y + c·z + ...) = 4F and (x + y + z + ...) = V, where finally then, n = 2E/F and p = 2E/V by definition, as throughout.

[118] J.M. Cooper, editor, *Plato: Complete Works*, 1st edition, Hackett Publishing Company, Indianapolis, IN, 1997.

[119] W.L. Bragg, *The Development of X-ray Analysis*, 1st edition, Dover Publications, Inc., Mineola, NY, 1975.

[120] L. Pauling, *The Nature of the Chemical Bond*, 3rd edition, Cornell University Press, Ithaca, NY, 1960.

[121] P. Duchowicz, M.J. Bucknum and E.A. Castro, *Journal of Mathematical Chemistry*, 41(2), 193, (2007).

[122] (a) M.J. Bucknum and E.A. Castro, *(MATCH) Commun. Math. Comp. Chem.*, 55, 57, (2006). (b) M.J. Bucknum and E.A. Castro, *Solid State Sciences*, in press, (2008).

[123] R.H. Baughman, H. Eckhardt and M. Kertesz, *Journal of Chemical Physics*, 87(11), 6687, (1987).

[124] (a) A.T. Balaban, C.C. Rentea and E. Ciupitu, *Rev. Roum. Chim.*, 13, 231, (1968), (b) H. Zhu, A.T. Balaban, D.J. Klein and T.P. Zivkovic, *Journal of Chemical Physics*, 101, 5281, (1994).

[125] V.H. Crespi, L.X. Benedict, M.L. Cohen and S.G. Louie, *Physical Review B*, 53, R13303, (1996).

[126] B. Wen, J. Zhao, D.Si, M.J. Bucknum and T.Li, *Diamond & Related Materials*, in press, (2008).

[127] M.J. Bucknum and E.A. Castro, *Journal of Chemical Theory & Computation*, 2(3), 775, (2006).

[128] A.T. Balaban, D.J. Klein and C.A. Folden, *Chemical Physics Letters*, 217, 266, (1994).

[129] H.R. Karfunkel and T. Dressler, *Journal of the American Chemical Society*, 114, 2285, (1992).

[130] M.J. Bucknum, *Chemistry Preprint Archive*, 2001(1), 75, (2001)

[131] J. Baggott, *Perfect Symmetry: The Accidental Discovery of Buckminsterfullerene,* 1st edition, Oxford University Press, Oxford, U.K., 1996.

[132] M.J. Bucknum and R. Hoffmann, *Journal of the American Chemical Society*, 116, 11456, (1994).

[133] M.J. Bucknum and E.A. Castro, *Journal of Theoretical & Computational Chemistry*, 5(2), 175, (2006).

[134] M.J. Bucknum and E.A. Castro, *Molecular Physics*, 103(20), 2707, (2005).

[135] M. O'Keeffe and B.G. Hyde, *Crystal Structures I. Patterns and Symmetry,* 1st edition, Mineralogical Society of America (M.S.A.), Washington, D.C., 1996.

[136] J. Waser and E.D. McClanahan, *Journal of Chemical Physics*, 19, 413, (1951).

[137] T. Siegrist, S.M. Zahurak, D.W. Murphy, R.S. Roth, *Nature*, 334, 231, (1988).

[138] M. Dincã, A. Dailly, Y.Liu, C.M. Brown, D.A. Neumann and J.R. Long, *Journal of the American Chemical Society*, 128, 16876, (2006).

[139] M.J. Bucknum and E.A. Castro, *Russian Journal of General Chemistry*, 76(2), 265, (2006).

[140] M.J. Bucknum and E.A. Castro, *Central European Journal of Chemistry (CEJC)*, 3(1), 169, (2005).

[141] M.J. Bucknum and E.A.Castro, *Journal of Mathematical Chemistry*, 42(3), 373, (2007).

[142] (a) P. Beckmann, *A History of π,* 1st edition, The Golem Press, New York, NY, 1971. (b) D. Blatner, The Joy of π, 1st edition, Walker Publishing Company, Inc., USA, 1997.

[143] E. Maor, e: *The Story of a Number*, 1st edition, Princeton University Press, Princeton, NJ, 1994.

[144] M.J. Bucknum, B. Wen and E.A. Castro, *Global Journal of Molecular Sciences*, in press, (2008).

[145] M. Livio, *The Golden Ratio: The Story of □ the World's most Astonishing Number,* 1st edition, New York, NY, Broadway Books, 2002.

INDEX

C

D

E

F

G

H

I

J

K

L

M

N

O

P

Q

R

S

T